Recent Advances in the 3D Physiological Human

Nadia Magnenat-Thalmann · Jian J. Zhang ·
David D. Feng
Editors

Recent Advances in the 3D Physiological Human

Springer

Editors
Dr. Nadia Magnenat-Thalmann
Université de Genève
MIRALab
Switzerland

Jian J. Zhang
Bournemouth University
United Kingdom

David D. Feng
University of Sydney
Australia

ISBN 978-1-4471-6895-9 ISBN 978-1-84882-565-9 (eBook)
DOI 10.1007/978-1-84882-565-9
Springer Dordrecht Heidelberg London New York

Printed on acid-free paper

Springer is part of Springer Science+Business Media (www.springer.com)

Preface

This book presents recent advances in the domain of the 3D Physiological Human that have been presented last December in the Workshop on 3D Physiological Human 2008. This workshop was funded by the "Third Cycle in Computer Science of Western Switzerland" named CUSO, the European project Focus K3D, and the European Marie Curie project 3D Anatomical Human (3DAH).

3D Physiological Human research is a very active field supported by many scientific projects. Many of them are funded by the European Union such as 3D Anatomical Human project (FP6-MRTN-CT-2006) and those present in the seventh framework program "Virtual Physiological Human" (FP7-ICT-2007-2). One of the main objectives of the research on 3D Physiological Human is to create patient-specific computer models for personalized health care. These models are used to simulate and hence better understand the human physiology and pathology.

A collection of scientific articles is proposed to highlight the necessity to exchange and disseminate novel ideas and techniques from a wide range of disciplines (computer graphics, biomechanics, knowledge representation, human–machine interface, etc.) associated with medical imaging, medical simulation, computer-assisted surgery, and 3D semantics. The emphasis is on technical novelty along with current and future applications for modeling and simulating the anatomical structures and functions of the human body.

The book is divided into three main parts: anatomical and physiological modeling, physically based simulation, and medical analysis and knowledge management. We would like to thank the program committee members for their work on reviewing papers.

We would also like to thank Jérôme Schmid and Caecilia Charbonnier from MIRALab-University of Geneva to have finalized the documents for the camera-ready edition of this book.

February 2009

Prof. Nadia Magnenat-Thalmann
MIRALab-University of Geneva, Switzerland
Prof. Jian Jun Zhang
Bournemouth University, UK
Prof. David Dagan Feng
Sydney University, Australia

Table of Contents

Part III Medical Analysis and Knowledge Management

Part I

Anatomical and Physiological Modeling

Musculoskeletal Simulation Model Generation from MRI Data Sets and Motion Capture Data

Jérôme Schmid[1], Anders Sandholm[2], François Chung[3], Daniel Thalmann[4], Hervé Delingette[5], and Nadia Magnenat-Thalmann[6]

Abstract Today computer models and computer simulations of the musculoskeletal system are widely used to study the mechanisms behind human gait and its disorders. The common way of creating musculoskeletal models is to use a generic musculoskeletal model based on data derived from anatomical and biomechanical studies of cadaverous specimens. To adapt this generic model to a specific subject, the usual approach is to scale it. This scaling has been reported to introduce several errors because it does not always account for subject-specific anatomical differences. As a result, a novel semi-automatic workflow is proposed that creates subject-specific musculoskeletal models from magnetic resonance imaging (MRI) data sets and motion capture data. Based on subject-specific medical data and a model-based automatic segmentation approach, an accurate modeling of the anatomy can be produced while avoiding the scaling operation. This anatomical model coupled with motion capture data, joint kinematics information, and muscle-tendon actuators is finally used to create a subject-specific musculoskeletal model.

[1] University of Geneva, MIRALab
schmid@miralab.unige.ch

[2] Ecole Polytechnique Fédérale de Lausanne, Virtual Reality Laboratory
anders.sandholm@epfl.com

[3] INRIA Sophia-Antipolis, Asclepios Research Team
francois.chung@sophia.inria.fr

[4] Ecole Polytechnique Fédérale de Lausanne, Virtual Reality Laboratory
daniel.thalmann@epfl.com

[5] INRIA Sophia-Antipolis, Asclepios Research Team
herve.delingette@sophia.inria.fr

[6] University of Geneva, MIRALab
thalmann@miralab.unige.ch

N. Magnenat-Thalmann et al. (eds.), *Recent Advances in the 3D Physiological Human*,
DOI 10.1007/978-1-84882-565-9_1,

Introduction

Today neuromuscular simulation has proved to be a valuable tool in understanding human movements, from gait research [19], diagnostic and treatment of patients to teaching/training new physicians. During the last years, several simulation platforms have been developed, both commercial [7, 10] and open source based [9], along with several models with different levels of details and complexity [29, 14].

The use of computer model/simulation has also allowed researchers and physicians to test what-if scenarios and even simulate different treatments before the physical treatment or surgery takes place. A person's gait cycle can be affected by many factors such as bone deformations [18] or static muscle contractors [25]. To simulate such patients with gait troubles or deformed skeleton, a full inspection of their anatomy should be first conducted. The most common neuromuscular models which are used today are all based on healthy, average-size adult male specimens [15]. Before the simulation of a person anatomy, the generic neuromuscular model has to be scaled to match the anatomy of the subject. The most two common ways are the isotropic scaling, where the whole model is scaled with the same scaling factor, and the anisotropic scaling, where each model part is scaled with an individual scaling factor. To determine these scaling factors, an initial motion capture is performed, where the subject stands still wearing 3D position markers placed at pre-defined anatomical positions. From these 3D markers, the length and position of each body segment can be computed to estimate the appropriate scaling factors to be applied on the generic musculoskeletal model. This scaling approach has several limitations and drawbacks. First, the surface-based recordings cannot fully account for the large anatomical differences that exist among individuals. Second, they do not tackle the issue of skin displacement with respect to the underlying bones. A slight misplacement of a marker can generate large differences in the simulated model. In patients with cerebral palsy, it has been shown that the length of moment arms can significantly differ between a scaled and a subject-specific model [23].

The scaling is not the only issue; the complexity of the musculoskeletal system should also be addressed. Various studies tried to set up anatomical models at various levels of complexity. Scheepers et al. [22] and Aubel and Thalmann [2] described approaches based on anatomical concepts and constraints but the methods used unrealistic simplifications. One of the major limitations was that both studies were not subject specific. More advanced models were presented like the works of Teran et al. [26] or Blemker and Delp [5] creating more complex anatomical models (e.g., fiber directions) to increase simulation accuracy. Although these studies relied on medical data sets offering the individualization of the modeling, they still require manual operations and corrections.

In this chapter, a new framework that combines data from motion capture and segmented MRI images is introduced to create a subject musculoskeletal model and therefore remove the initial scaling step. The anatomical modeling pipeline uses first a stage of interactive modeling to create what is referred to as a *generic anatomical model*. This modeling needs to be performed only once and it considers anatomical constraints. Then the generic anatomical model is used in an automatic procedure that can be applied on any data set to perform the individualization. The proposed approach can thus relieve the user from tedious interactive modeling. Finally, the anatomical model is combined with additional information produced from a reference motion capture frame to generate the complete musculoskeletal model in a semi-automatic fashion. If needed, a user interface is provided to alter/refine the full model. This model can then be used to simulate various motions.

Musculoskeletal Modeling Pipeline

When generating a musculoskeletal model, several sources of information and data have to be combined in a single simulation setup. The overall workflow, depicted in Fig. 1, is divided up into two phases, *a* and *b*. In phase *a*, an individualized musculoskeletal model describing the subject is created, this model can be subsequently used in phase *b* where a neuromuscular simulation is carried out to analyze a desired motion.

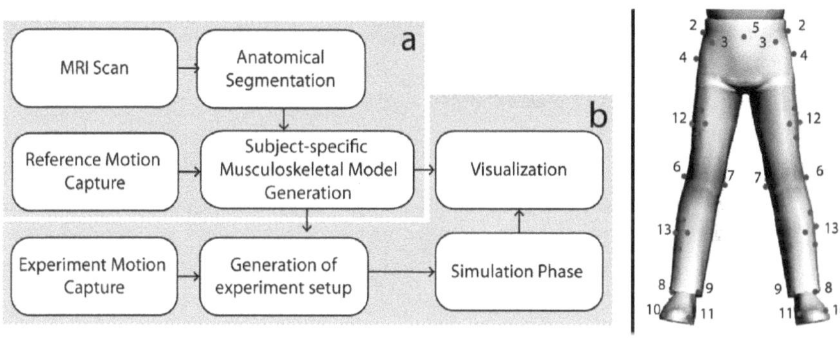

Fig. 1. *Left*: Workflow of subject-specific neuromuscular model generation and simulation. Phase *a* generates the subject-specific neuromuscular model; in phase *b* simulation and visualization are performed. *Right*: Reference (*blue*) and skin (*red*) markers position overview, marker numbers correspond with Table 1. Only the lower limb markers are shown.

In the first phase, a data acquisition process takes place. The lower limbs of the subject are scanned in a MRI device to produce data sets that are later on segmented (see corresponding section) in various anatomical components (e.g., muscles, bones). Motion capture sessions are also performed to collect information during a standing reference motion capture and one or several experiment motion captures. The latter motion capture is coupled with synchronized EMG acquisitions and is used to simulate the motion that we want to study. By combining the 3D models produced from the segmentation with the reference motion capture data, the subject-specific musculoskeletal model is generated. This model is then used in the second phase, along with the experiment motion capture data to simulate and analyze a specific motion.

In developing this framework, the OpenSim platform was chosen as the neuromuscular simulation engine, mainly because it is based on an open-source license and it supports several tools for model editing, fast simulation, and advanced visualization features. OpenSim also supports a large collection of simulation possibilities such as inverse kinematics, muscular control, and forward dynamics, whose output parameters include joint forces, muscle-induced acceleration, and muscle power.

Data Acquisition

Markers Setup for MRI and Motion Capture

To create the musculoskeletal model, two marker sets are utilized. One set of *skin markers* is placed on muscle tissue while the other set of *reference markers* is placed on bone features, see Table 1 and right side of Fig. 1.

Table 1 Reference and skin markers

	Placement	Type	Marker set
1	Acromioclavicular	Single marker	Reference marker
2	Iliac crest	Single marker	Reference marker
3	ASIS	Single marker	Skin marker
4	Greater trochanter	Single marker	Reference marker
5	Sacrum	Single marker	Skin marker
6	Lateral knee	Single marker	Reference marker
7	Medial knee	Single marker	Reference marker
8	Lateral ankle	Single marker	Reference marker
9	Medial ankle	Single marker	Reference marker
10	5th metatarsal head	Single marker	Skin marker
11	1st metatarsal head	Single marker	Skin marker
12	Thigh	Cluster set	Skin marker
13	Shank	Cluster set	Skin marker

The reference markers are located on easy-to-find bone features and are used during both the MRI acquisition and the motion capture. During motion capture, reflection markers are used, while markers filled with contrast agent are used during the MRI acquisition. The reference markers are used to put in correspondence the segmented and the motion capture data. Due to the fact that the reference markers are placed by hand at two different acquisitions, a small difference in placement can occur. However, this error is smaller than the error produced by skin movements during acquisition.

MRI Acquisition

In close collaboration with radiologists, an adequate protocol for the imaging of soft and bony tissues was defined: Axial 2D T1 Turbo Spin Echo, TR/TE = 578/18 ms, FOV/FA = 40 cm/90°, matrix/resolution = 512×512/0.78×0.78 mm. To acquire a full lower limb, two different sessions took place and during each of them three acquisitions were performed. Each acquisition was based on the same protocol, but with varying slice thickness (e.g., higher thickness for long bones) to speed up the acquisition process. First session consisted in hip (thickness: 2 mm), thigh (10 mm), and knee (4 mm); second session in knee (2 mm), leg (10 mm), and foot (4 mm). All acquisitions of a session were merged in a unique volume, and the two session volumes were registered together (thanks to an appropriate knee overlap). After cropping and resampling operations, two data sets have been created with the following size (resolution): 202 × 398 × 595 (0.78 × 0.78 × 2.46 mm) and 151 × 213 × 582 (0.68 × 0.68 × 2.5 mm) for the first and second data sets, respectively. All acquisitions were performed at the University Hospital of Geneva on a 1.5 T MRI device (Philips Medical Systems).

Motion Capture Acquisition

For the recording of the reference and experiment motion capture data, an optical motion capture system of eight ProReflex Qualisys (www.qualisys.com) video cameras was used. Prior to the session, reflective skin markers were fitted to the lower limbs according to Table 1. Subjects were also equipped with electromyography (EMG) sensors from Noraxon (www.noraxon.com) to capture electrical activity in eight large lower limb muscles, see Table 2. During motion acquisition, two AMTI (www.amti.biz) force platforms were used to capture ground reaction force. Marker trajectories, EMG, and force plate data were acquired, labeled, and exported by using Qualisys Track Manager to C3D files.

Table 2 Electromyography electrodes placement

No	Muscle	
1	Tibialis anterior	with additional ground marker
2	Gastrocnemius	
3	Vastus medialis	
4	Soleus	
5	Vastus lateralis	
6	Rectus femoris	
7	Biceps femoris	
8	Gluteus maximus	

Segmentation

In order to model the various structures of interest with a high fidelity, a segmentation procedure is applied on medical data sets of the subject. The usage of manual segmentation is twofold. First, it serves as a basis to automatic methods by providing (i) some prior knowledge (e.g., internal forces in deformable models that enforce the consistency with an a priori of a shape [8, 12]) or (ii) training material (e.g., methods based on principal component analysis (PCA) [6, 24] or classification methods like K-means [13] and Fuzzy C-means [4]). In general, more than one model is needed to produce satisfactory results. Second, it provides a ground truth for experiments. Segmentations generated by an automatic method can thus be compared with those obtained manually. In our case, manual segmentation has been performed by a medical expert, which allows us to validate it. In the following, our methodology for manual segmentation will be presented and then a semi-automatic segmentation method will be depicted.

Generic Anatomical Model Construction

Manual Delineation

The manual delineation of the bones and muscles has been performed by a medical expert (see Fig. 2a). For each structure, a binary image was produced and then processed by the Marching Cubes algorithm [17] to generate 2-simplex meshes [8] representing the structures of interest. However, the generated meshes surface appeared not to be smooth enough. This is mostly explained by the data set resolution (especially the large interslice distance) and the ubiquitous error made during a manual segmentation, which remains a subjective task. Ideally, the resolution should be as high as possible to provide detailed images and to avoid big jumps between consecutive MRI slices. This is not always feasible (e.g., device limitations, acquisition time restrictions). Furthermore, a lower resolution is preferred to reduce the memory footprint which results in a speed-up automatic algorithm. Since muscles are known to be anatomically smooth, the objective is to find a solution that makes them appear so. The technique we used consists first in refining the meshes and then in applying internal constraints on them. More precisely, we applied forces on the meshes so that their rigidity is the maximum. For that, we used the simplex angle continuity or C_2 constraint [8] used for shape recovery and which is defined as an average of the simplex angles at neighborhood vertices. Since we want to smooth our meshes in a large scale, we defined a great neighborhood size which intuitively corresponds to the notion of high rigidity for the mesh. The result was that most pits and bumps present on the meshes surface disappeared throughout the deformation process (see Fig. 2c). The smoothing based on C_2 constraint has also the advantage to reduce the shrinking effect obtained with standard Laplacian smoothing.

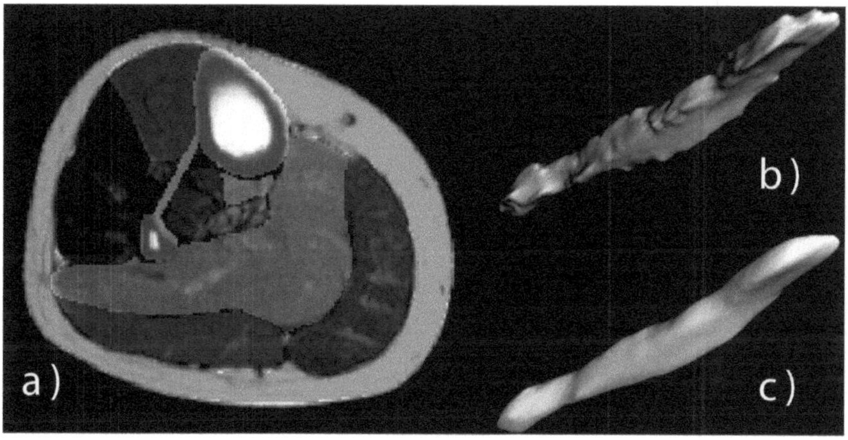

Fig. 2. (**a**) Axial view of the leg with manual delineation in half transparency; (**b**), (**c**) result of the smoothing procedure on the flexor digitorum longus muscle: (**b**) mesh surface generated from the manual delineation and (**c**) mesh surface smoothed using the simplex angle continuity or C_2 constraint.

Tendons

Although the muscles are difficult to delineate manually, the tendons are easier to detect in the MRI. Indeed, they have a uniform appearance in the images (tendons appear as dark structures with respect to muscle tissue). An automatic method based on maximum intensity ridge tracing [3, 20] was first used. Such method relies on the assumption that tendons are roughly tubular structures, which is especially the case for the leg tendons. These tracing methods have proved to be robust to noise. They also provide a simple way to model a structure as a series of centerline points (a position and a radius). Furthermore, such representation could be reused in another automatic method. However, this method may not always segment correctly or entirely the tendons because in some regions their intensity is corrupted by too strong artifacts (e.g., noise, adjacent structures, partial volume effect). As a result, tendons are difficult to follow through the slices. Moreover, some tendons are so close (e.g., fibularis longus and brevis tendons) that it is not possible to distinguish them without some prior knowledge.

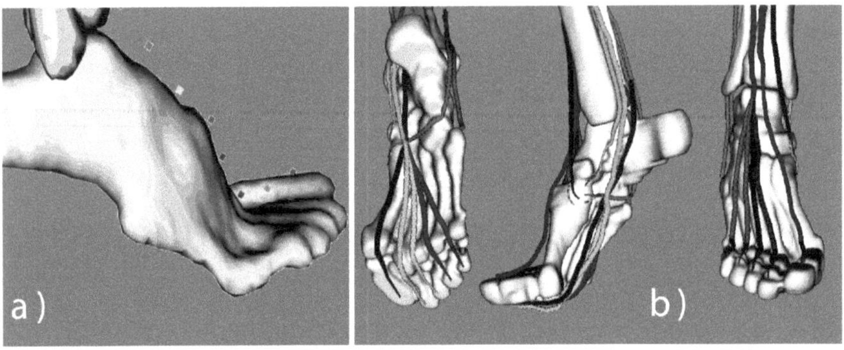

Fig. 3. *Left*: manual placement of centerline points for the Extensor hallucis longus tendon starting from the eponymous muscle until its attachment above the big toe. *Right*: results on various foot tendons.

To cope with these problems, missing centerline points were manually placed (Fig. 3a) on high curvature points of tendons using an editing tool, such as CardioViz3D [27]. Eventually, when the centerlines are complete (thanks to the automatic or manual modeling), simplex surfaces of tendons are automatically created. Figure 3b depicts some of the results for the foot tendons complex modeling.

Attachments

Anatomically, an attachment is defined as the linking region between two structures. For instance, tendons are usually attached to bones. The tendon attachments are thus defined as the tendon tissues sharing a common region with the bone tissues. Comparatively, it means that the meshes modeling these structures should be stuck together which is not the case a priori. Indeed, as accurate as the segmentation may be, the different generated meshes are not guaranteed to be attached together, especially after the smoothing procedure. Our solution is to manually place splines to define attachment regions as described in [11]. This provides a simple but efficient way to model attachment regions. Meshes are then deformed until they are stuck.

Interpenetrations Removal

The smoothing and attachment procedures previously described can create nonrealistic interpenetrations between meshes. In order to remove them, a geometrical post-processing is applied. Let us consider the surfaces S_1 and S_2 in Fig. 4 that are interpenetrating each other. The aim is to move the colliding points of each surface, located in the penetration volume, to reach a non-interpenetrating state.

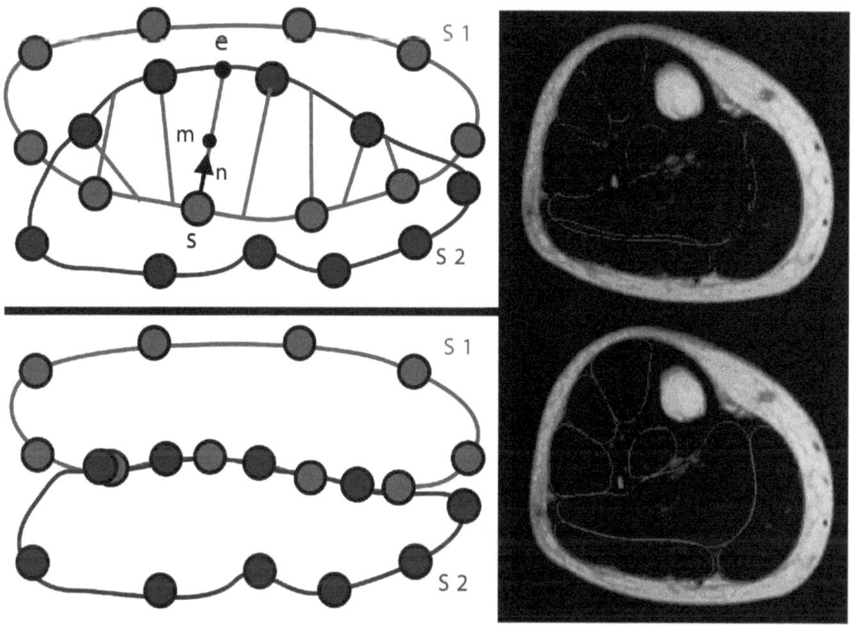

Fig. 4. *Top-left*: surfaces S1 and S2 are interpenetrating each other. *Bottom-left*: interpenetrations removal results. *Right*: results in real cases. It can be seen that muscles are not diffused through the neighbors anymore.

For a point $S \in S_1$ with inner normal n, the line passing through s and directed by n intersects S_2 in e. This point e must be inside S_1 to be valid. The new position m of the point s is then chosen as

$$\begin{cases} m = \{x \in [s,e] / f(m) \text{ is minimal}\} \\ f(x) = (l_1 * d_1(x) - l_2 * d_2(x))^2 \end{cases} \quad (1)$$

where $d_i(x)$ depicts the signed distance from x to the surface S_i and l_i is a rigidity parameter. For instance, a configuration with $l_1 << l_2$ will lead to final points closer to surface S_2. This can be useful to resolve penetrations between a bone (very stiff material) and a muscle. The presented technique was inspired from [16], with the distinction that point e is not defined as the projection of s on S_2. Indeed, by using this projection, the point m does usually not correspond to the new position of s in the noncolliding state. By using instead the normal direction n, the point m will provide a better approximation. As a result, this technique avoids the usage of a resampling scheme as described in [16], points just need to be moved to their new positions. A *perfect* contact between the surfaces is nevertheless not ensured but the results are satisfactory when meshes resolution is not too low. Finally, implicit surfaces [1] are not used since the hypothesis of vicinity to surfaces may be invalid (i.e., in case of big interpenetrations). To speed up the process, precomputed signed distance maps can be used as well as the efficient golden section search technique [21] to minimize f.

Semi-automatic Segmentation

The result of the previous section is a generic anatomical models collection of the various soft (muscles, tendons) and bony structures, from which various levels of details (LODs) are computed (four for bones and three for soft structures). The objective now is to re-exploit these data as prior models in a semi-automatic method. This method could then be applied on any similar data set coming from other subjects. The proposed method is mostly based on the deformable models method of Gilles et al. [12]. Mesh vertices are considered as lumped mass particles subjected to internal/external forces and evolving under the Newtonian law of motion.

Internal forces regulate shape evolution by enforcing constraints [8, 12] on the surface regularity (smoothness, curvature) and on the shape configuration. Shape constraints are simply enforced by predefining their simplex parameters. We added other priors [24] based on PCAs of training shapes and on the modeling of local deformations as a Gauss–Markov random process. External forces attract meshes toward anatomical boundaries by using image information (e.g., intensity, gradient) and impose non-penetration constraints. The attachments use mass modification methods to constrain vertices, while collision handling techniques prevent mesh penetrations through penalty forces.

The Newton equation relates forces to particle state. The resulting differential equations system is time discretized and solved by a stable implicit integration scheme [28]. The different mesh LODs are used in a multilevel framework using a force propagation mechanism. This strategy confers robustness and accuracy to the segmentation method. Conducted experiments [12, 24] reported an average accuracy of 1.5 mm for bone and muscle segmentation. The described segmentation procedure is equivalent to a model to image registration. It is semi-automatic because generic anatomical models are initialized manually by placing landmarks. The power of using and registering generic shapes is that exact geometric correspondences are obtained (i.e., morphological features have the same vertex indexes across individual models, like attachments for instance).

Anatomical Modeling Validation

The proposed methods for correction (smoothing and interpenetrations removal) and the semi-automatic algorithms are still prone to errors. Indeed, they cannot always guarantee that the segmentation will perfectly delineate the structures of interest. To correct the most significant errors, a medical expert can define manually some constraints by placing points (e.g., points on the organ boundaries) in the MRI data sets. Forces are then applied on the meshes to account the expert constraints [11]. Similarly, forces based on image information (e.g., gradient) can be used to perform some local corrections. This approach gives nice results but as a general remark, one should avoid to place too many constraints or be tempted to blindly trust the image information. Indeed, the placement of constraints can present the same pitfalls as a manual delimitation: it is difficult to assess the organs transition between slices and ensure meshes smoothness. Consequently, it is essential to find a trade-off between reconstructed models quality and exact segmentation, since images cannot be totally trusted. Most importantly, the purpose of the segmentation is to provide *models* that must be reusable in our pipeline or in other studies.

Musculoskeletal Model Generation

The generation of a subject-specific musculoskeletal model can be divided into two parts. First, the anatomical segmented 3D objects have to be aligned with the reference motion capture data and second, the segmented objects have to be extended with additional information to create a complete musculoskeletal model.

Data Alignment

The first step in generating a subject-specific musculoskeletal model is to align the segmented data with the motion capture reference frame, as depicted in the first step of the workflow (see Fig. 1). Doing this, the correct placement of the motion capture skin markers can be included into the model. This alignment is also performed to prepare the anatomical 3D objects for visualization purpose.

The alignment starts by reading the reference markers from the imported C3D file along with the reference markers from the segmented MRI images. For each bone segment (pelvis, femur, and tibia), a rigid transformation matrix is calculated so that each segment can be correctly aligned with the motion capture data. To translate each bone segment, its corresponding markers are used, see Table 3. Currently, the marker extraction in the MRI images is performed manually but an automatic method could be achieved by using standard thresholding and mathematical morphology operators. The combination of various subject-specific data avoids performing any error-prone scaling.

By using each bone segment transformation, the skin markers defined in the reference motion capture file can be similarly expressed in the MRI coordinate space. For example, the skin marker thigh (see Table 1) is transformed using the femur matrices (see Table 3).

Table 3 Markers used to transform each bone segment

Bone segment	Reference motion capture marker
Pelvis	Left and right side markers of iliac crest, left and right side markers of greater trochanter
Femur	Greater trochanter, lateral knee, medial knee
Tibia	Lateral knee, medial knee, lateral ankle, medial ankle

Model Generation

When the initial step of alignment is completed, the model consists of the 3D segmented objects and of motion capture markers. To generate a complete musculoskeletal model, information about muscles and joints has to be included. From the segmented 3D objects, each joint center is calculated. This joint center is used as the center point for the moment force generated by the muscles. Then for each joint, a kinematic function is defined to describe the kinematic behavior of each joint. By using this function and a given joint angle, the relative attachment point of the joints can be determined. Currently, these data are derived from Delp et al. [10]. To calculate the force that each muscle and its tendons can generate, each muscle is expressed into a set of one or more muscle-tendon actuators. These muscle-tendon actuators stretch between the muscles insertion points either as a straight line or via a wrapping point. At a wrapping point, the actuator changes direction to better describe the shape of the muscle as shown in Fig. 5. Wrapping points are also used to prevent muscle-tendon actuators to collide and enter a bone segment. If such a collision occurs, the wrapping becomes a fixed point and the actuator "wraps" around the bone, see Fig. 5b, c. The muscle-tendon actuator (Fig. 5a) is connected to each joint by using the tendon data from the segmentation process (see section “Attachments”), and the framework supports manual correction of both insertion and wrapping points. Currently, the description for all muscle tendons and wrapping points is derived from Delp et al. [10] and OpenSim [9]. Eventually, the model is ready to be exported as a set of XML files in the OpenSim format.

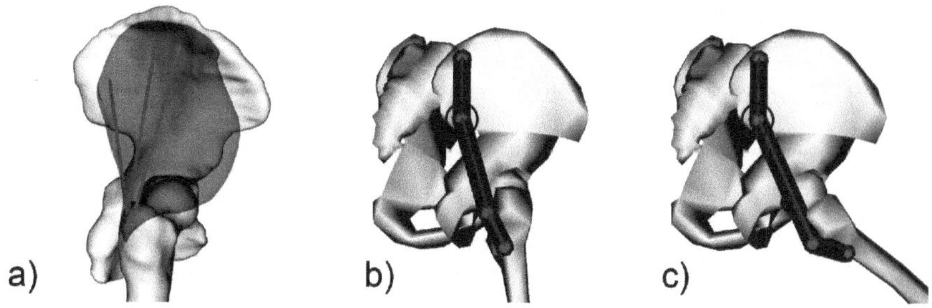

Fig. 5. (**a**) Visualization of muscle-tendon actuators in gluteus medius model; (**b**), (**c**) Muscle-tendon actuator representing Gluteus maximus; images (**b**) and (**c**) show the influence of wrapping points.

When a subject-specific simulation is carried out, the generated musculoskeletal model is loaded along with the experiment motion capture data. The model will now match the motion capture data, so no initial scaling has to be carried out. The only initial step before starting a simulation is to align each joint angle in the model with the corresponding joint angle in the first frame of the experiment motion capture. This is done using an inverse kinematic calculation which is performed by the OpenSim framework. The musculoskeletal model should now match the first frame of the motion capture data and the desired simulation can be executed.

Results and Discussion

Anatomical Modeling Results

To illustrate the anatomical modeling, we applied our methodology on data sets produced from a female and healthy subject (age 24, height 1.68 m, weight 58 kg) acquisition. The institutional medical-ethical committee approved the study and the subject gave written informed consent. The manual modeling used to construct the generic anatomical model is applied on the leg area (knee to foot) while the automatic method is performed on the hip–thigh–knee area. The automatic method relies on generic models of the thigh previously generated from a similar manual interactive method. Results are visible in Fig. 6a–d.

Musculoskeletal Modeling Results

Effort has been put in devising a framework to set up a musculoskeletal model from subject-specific data. Currently, this workflow has only been used on one subject, some results are visible in Fig. 6e. Therefore, no quantitative comparison between scaled and subject–specific simulation is available. However, the initial step of scaling has been avoided and this should therefore remove a source of error in the musculoskeletal simulation. The proposed approach also supports the generation of musculoskeletal models from subjects with no conventional bone structures (e.g., pathological bones), where scaling would not have been possible since the underlying bone structure can be hard to determine from the surface markers/measurements.

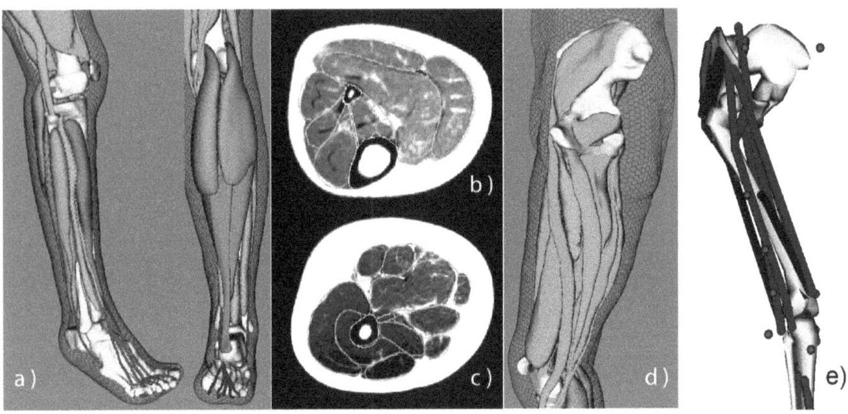

Fig. 6. Anatomical and musculoskeletal modeling results: (**a**) Lateral and posterior view of the leg based on manual modeling; (**b**) slice with segmentation contours overlay of the leg; (**c**) slice with segmentation contours overlay of the thigh; (**d**) lateral view of the thigh based on automatic segmentation; (**e**) musculoskeletal simulation in OpenSim showing bone surface, muscle-tendon actuators, and motion capture markers.

Conclusion and Future Work

The workflow outlined in this chapter allows the user to create a subject-specific musculoskeletal model based on MRI and motion capture acquisitions. The process is described as a semi-interactive approach where the segmentation is performed using a generic anatomical model. The segmented data are then extended with joint and muscular information so that a full musculoskeletal model is created. Currently, joint kinematics and tendon-muscular actuator information are derived from recognized literature. When the specific musculoskeletal model of the subject is generated, it can be used with any subject motion capture data to produce a desired simulation. The versatility of the framework is one of its qualities as it is not anymore necessary to rely on a standardized healthy male musculoskeletal model. It should be hence possible to simulate patients with disabilities as long as the acquired data reflect their pathology (e.g., organ shapes, abnormal EMG patterns, not conventional motions). Moreover, the framework being coupled with the OpenSim platform, a wide variety of measurements and simulation scenarios are available.

Yet a lot of work remains to fully validate the framework. Especially, more experiments must be carried out to compare the standard approach based on scaling and the proposed one. Accent will be put in studying more carefully the joint kinematics and the muscle-tendon actuators. The proposed workflow will also have to study, from a simulation viewpoint, the differences that exist among individuals such as gender, size, and age. The approach will then be used in a medical context where some specific motions on pathological subjects will be examined. We hope that it will give some highlights into the mechanisms of pathologies and their treatments.

Acknowledgments This work is supported by the 3D Anatomical Human project (MRTN-CT-2006-035763) funded by the European Union.

References

1. Alexa M, Adamson A (2004) On normals and projection operators for surfaces defined by pointsets. In: M. Alexa, M. Gross, H. Pfister, S. Rusinkiewicz (eds.) Proc. Eurographics Symposium on Point-based Graphics, pp. 149–156
2. Aubel A, Thalmann D (2001) Interactive modeling of the human musculature. In: Hyeong-Seok Ko (ed.) Proc. of Computer Animation, pp. 167–174
3. Aylward S, Bullitt E (2002) Initialization, noise, singularities, and scale in height ridge traversal for tubular object centerline extraction. IEEE Transactions on Medical Imaging 21(2):61–75
4. Bezdek J (1973) Fuzzy mathematics in pattern classification. PhD Thesis, Cornell University, Ithaca, NY
5. Blemker S, Delp S (2005) Three-dimensional representation of complex muscle architectures and geometries. Annals of Biomedical Engineering 33:661–673
6. Cootes T, Hill A, Taylor C, Haslam J (1994) The use of active shape models for locating structures in medical images. Image and Vision Computing 12(6):355–366
7. Damsgaard M, Rasmussen J, Christensen ST, Surma E, de Zee M (2004) Analysis of musculoskeletal systems in the anybody modeling system. Simulation Modelling Practice and Theory 14(8):1100–1111

8. Delingette H (1999) General object reconstruction based on simplex meshes. International Journal of Computer Vision 32(2):111–146
9. Delp SL, Anderson FC, Arnold AS, Loan P, Habib A, John CT (2007) Opensim: Opensource software to create and analyze dynamic simulations of movement. IEEE Transactions on Biomedical Engineering 54(11):1940–1950
10. Delp SL, Loan JP, Hoy MG, Zajac FE, Topp EL, Rosen JM (1990) An interactive graphics-based model of the lower extremity to study orthopaedic surgical procedures. IEEE Transactions on Biomedical Engineering 37(8):757–167
11. Gilles B (2007) Anatomical and kinematical modelling of the musculoskeletal system from mri. PhD Thesis, University of Geneva
12. Gilles B, Moccozet L, Magnenat-Thalmann N (2006) Anatomical modelling of the musculoskeletal system from mri. In: R. Larsen, M. Nielsen, J. Sporring (eds.) MICCAI 2006. LNCS, vol. 4190, pp. 289–296
13. Hartigan J, Wong M (1979) A K-means clustering algorithm. Journal of the Royal Statistical Society Series C: Applied Statistics 28:100–108
14. Holzbaur K, Murray W, Delp S (2005) A model of the upper extremity for simulating musculoskeletal surgery and analyzing neuromuscular control. Annals of Biomedical Engineering 33(6):829–840
15. Horsman MK, Koopman H, van der Helm F, Prose L.P, Veeger H (2007) Morphological muscle and joint parameters for musculoskeletal modelling of the lower extremity. Clinical Biomechanics 22:239–247
16. Keiser R, Müller M, Heidelberger B, Teschner M, Gross M (2004) Contact handling for deformable point-based objects. In: B. Girod, M. Magnor, H.P. Seidel (eds.) Proc. of Vision, Modeling, Visualization '04 (VMV'04), pp. 315–322
17. Lorensen W, Cline H (1987) Marching cubes: A high resolution 3D surface construction algorithm. In: M.C. Stone (ed.) Proc. of the 14th Annual Conference on Computer Graphics and Interactive Techniques, pp. 163–169. ACM New York, NY, USA
18. Metaxiotis D, Acclesa W, Siebela A, Doederleina L (2000) Hip deformities in walking patients with cerebral palsy. Gait & Posture 11(2):86–91
19. Piazza S, Delp S (1996) The influence of muscles on knee flexion during the swing phase of gait. Journal of Biomechanics 29(6):723–733
20. Pock T, Janko C, Beichel R, Bischof H (2005) Multiscale medialness for robust segmentation of 3d tubular structures. In: A. Hanbury, H. Bischof (eds.) Proc. 10th Computer Vision Winter Workshop, pp. 93–102
21. Press WH, Flannery BP, Teukolosky SA, Vetterling WT (1992) Numerical Recipes in C, Second ed. Cambridge University Press, Cambridge
22. Scheepers F, Parent R, Carlson W, May S (1997) Anatomy-based modeling of the human musculature. In: SIGGRAPH'97, pp. 163–172. ACM Press/Addison-Wesley Publishing Co. New York, NY, USA
23. Scheysa L, Campenhoutc AV, Spaepenb A, Suetensa P, Jonkers I (2008) Personalized mr-based musculoskeletal models compared to rescaled generic models in the presence of increased femoral anteversion: Effect on hip moment arm lengths. Gait & Posture 28(3):358–365
24. Schmid J, Magnenat-Thalmann N (2008) MRI bone segmentation using deformable models and shape priors. In: D. Metaxas, L. Axel, G. Szekely, G. Fichtinger (eds.) MICCAI 2008, Part I. LNCS, vol. 5241, pp. 119–126. Springer-Verlag, Berlin, Heidelberg
25. Tardieu G, Tardieu C (1987) Mechanical evaluation and conservative correction of limb joint contractures. Clinical Orthopaedics & Related Research 219:63–69
26. Teran J, Sifakis E, Blemker S, Ng-Thow-Hing V, Lau C, Fedkiw R (2005) Creating and simulating skeletal muscle from the visible human data set. IEEE Transactions on Visualization and Computer Graphics 11:317–328
27. Toussaint N, Mansi T, Delingette H, Ayache N, Sermesant M (2008) An integrated platform for dynamic cardiac simulation and image processing: Application to personalised Tetralogy

of Fallot simulation. In: C.P. Botha, G. Kindlmann, W.J. Niessen, and B. Preim Proc. Eurographics Workshop on Visual Computing for Biomedicine (VCBM)
28. Volino P, Magnenat-Thalmann N (2005) Implicit midpoint integration and adaptive damping for efficient cloth simulation. Journal of Computer Animation and Virtual World 16(3–4): 163–175
29. Zhang LQ, Wang G (2001) Dynamic and static control of the human knee joint in abduction–adduction. Journal of Biomechanics 34(9):1107–1115

GeomCell
Design of Cell Geometry

Július Parulek[1], Miloš Šrámek[2], and Ivan Zahradník[3]

Abstract From the viewpoint of geometry, the structure of living cells is given by the 3D organization of their numerous intracellular organelles of various sizes, shapes, and locations. To understand them in their complexity, realistic computer models of cells may be instrumental and may moreover serve for virtual experiments and simulations of various kinds. We present a modeling concept based on the theory of implicit surfaces that allows for creation of a realistic infrastructure of the microworld of muscle cells. Creation of such models, consisting of hundreds or even thousands of organelles by means of traditional interactive techniques would, however, require unacceptably long time. Therefore, the whole model as well as each implicit object is created in an automatic process, guided by local and global geometric and statistic parameters. To accomplish this, we designed an XML-based cell modeling language. Further, the modeling system is supplemented by post-processing tools for model polygonization and voxelization, and, owing to high computational demands, was implemented in a grid environment.

Introduction

Recent progress in medical and biological sciences, especially in the area of health care, and the growth of computational power of up-to-date computers ask for development of virtual environments, which would provide users with geometrical modeling tools capable of creating 3D models of biological structures. Such virtual environments would allow researchers and doctors to grasp and

[1] Institute of Molecular Physiology and Genetics, Slovak Academy of Sciences, Slovakia
Faculty of Mathematics, Physics and Informatics, Comenius University, Slovakia
julius.parulek@savba.sk

[2] Faculty of Mathematics, Physics and Informatics, Comenius University, Slovakia
Commission for Scientific Visualization, Austrian Academy of Sciences, Austria
milos.sramek@oeaw.ac.at

[3] Institute of Molecular Physiology and Genetics, Slovak Academy of Sciences, Slovakia
ivan.zahradnik@savba.sk

N. Magnenat-Thalmann et al. (eds.), *Recent Advances in the 3D Physiological Human*,
DOI 10.1007/978-1-84882-565-9_2,

communicate very complex features of biological objects and to provide quality training and learning base for future students. Additionally, such models can be used for simulation or optimization. Numerous interdisciplinary research initiatives are generating excellent research results with regard to modeling, simulation, and visualization of human anatomy and physiology. These research initiatives focus on different biological levels [22], such as molecular and cellular levels, tissue [7] and organ levels [24, 25], and system and human (organism) levels. The GeomCell project, presented in this chapter, was initiated with the aim to exploit the possibilities of modern computer hardware and computer graphics tools in order to provide the biologist with an environment for virtual cell modeling based on a precise geometric background.

In this environment, the task is to equip the end users with tools that would enable them to build arbitrarily complex cell models and to verify these models according to statistical quantification attributes obtained from real measurements. These models can later be used by biologists to perform virtual experiments. A typical cell consists of hundreds or even thousands of various organelles and can be represented only by heterogeneous models composed of numerous objects. Thus, creation of such models organelle-by-organelle, using the traditional interactive techniques, would be too laborious and thus would require unacceptably long time. Therefore, in our approach, the whole cell model is created in an automated process, which, moreover, enables to reflect the statistical uncertainty of organelle occurrence, as well as of its shape and concrete dimensions.

According to the requirement of automated model creation, we have developed an XML-based model description language (MDL) [10], where free parameters are specified in a probabilistic sense. Each MDL configuration defines organelles in the form of their occurrence probability, typical sizes, and shape descriptors, including their stochastic variability. Therefore, as a benefit, a single MDL configuration can guide generation of numerous distinct, albeit statistically equivalent, models (Fig. 1).

The resultant models must fulfill requirements regarding both visual fidelity and quantitative relationships between volumes and areas of the individual organelles. Model visualization (Fig. 1) can reveal possible structural inconsistencies, and computation of the volume and surface densities (VSDs) of model organelles can reveal possible statistical inconsistencies. Visualization is a time-demanding process, because our cell models are composed of numerous objects represented by implicit functions, fast rendering of which is still a challenge [2]. VSDs of the modeled organelles have to be evaluated in order to compare them to the required parameters estimated from real cell data by stereological methods [26]. Such evaluation requires numerical computation of volume and surface integrals over each organelle in the model.

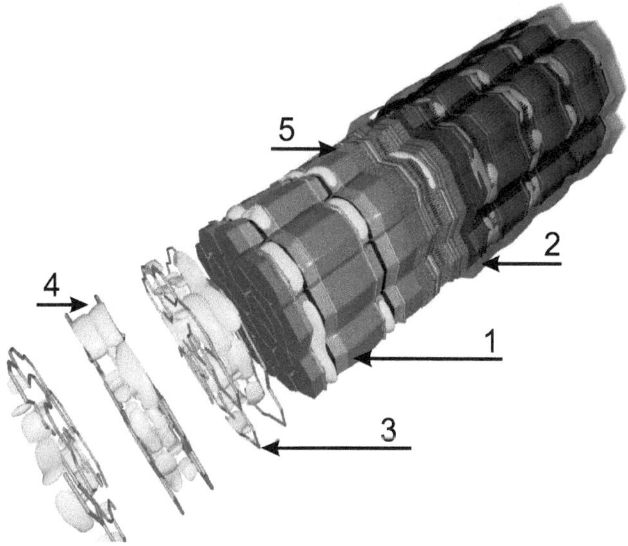

Fig. 1. Visualization of a cell model comprising the following organelles. 1—myofibrils, 2—transparent sarcolemma, 3—t-tubules, 4—mitochondria, and 5—sarcoplasmatic reticulum. Myofibrils and sarcolemma are clipped off by a plane perpendicular to the longitudinal axis of the cell. For better clarity, myofibrils were removed from the three bottom segments of the model.

According to the stochastic nature of the cell creation process, however, it is often the case that the individual created models do not meet all the user requirements. Since it is not possible to ensure these global properties in modeling of individual organelles, an iterative process is required, in which its input parameters are adaptively modified. In order to obtain a model with stereologic parameters (VSDs) which correspond to those of a real cell, construction and evaluation of hundreds of intermediate models may be required in this iterative process. Obviously, both model visualization and evaluation are extensive tasks that ask for computational power from the area of parallelization. Therefore, in our work, we turn to grid environments that, apart from parallelization of computations themselves, offer us the possibility to implement additional tools fulfilling custom user demands (e.g., model visualization), and also facilitate user requests through the web interface and the GUI portal.

In this chapter, we aim at the first building blocks of this project—design and representation of geometry of cell structures in order to represent a real muscle cell. The virtual 3D cell models are constructed and represented by means of implicit modeling techniques. Each organelle's outer membrane is represented by the 0-level isosurface of an implicit function. It is important to know that the positive aspect of implicits lies in the possibility to create organic-like shapes easily. Note that microworld cellular structures are mainly round-shaped objects without sharp corners. Modeling of such shapes can be achieved by combination of round primitives and proper operations applied on them. For instance, a special class of blend operations [18, 19] is a suitable device for creating smooth transitions between input round primitives such as algebraic primitives (quadrics) and skeleton-based primitives (distance-based objects, convolution surfaces). In addition, implicit surfaces also simplify point classification, collision detection, and object deformation, which are useful features in modeling of multi-organelle cells.

The remainder of the chapter is organized as follows: In the next section, we present previous work on muscle cell modeling. In more detail, we explain the design of cell geometry (third section) and the technique that is used in mitochondria modeling (fourth section). Then, in the fifth section we briefly introduce the cell creation process based on MDL and further, in the sixth section, we describe the grid implementation of the whole environment. The future work is introduced in the seventh section and we conclude in the eighth section.

Previous Work on Muscle Cell Models

In our first studies [16] we proposed methods that enabled us to create interactive tools providing users with a possibility of constructing organelles with lower complexity. While such models were useful for testing of specific stereologic methods, the modeling system required a large number of input parameters to be specified by an operator, which limited modeling to simple organelles (myofibrils, sarcolemma, mitochondria, and t-tubules) and short muscle cell segments. The next step was aimed at automation of the cell creation process [15].

Nevertheless, to be able to move toward this automation, we first had to develop an underlying implicit modeling environment fulfilling the demands of cell modeling, definition and storage of arbitrary implicit models, their visualization, polygonization, voxelization, and volume and surface area computation.

XISL—Implicit Modeling Environment

The XISL (XML-based scripting of Implicit Surfaces [13]) package assists developers in construction of arbitrary implicit models. The implicit models are described in declarative text files by means of the extensible markup language (XML). Each implicit function class (a primitive, an operation, etc.) is defined by its appropriate XISL tag(s). This ensures clear and self-explanatory notation of complex implicits (Fig. 2).

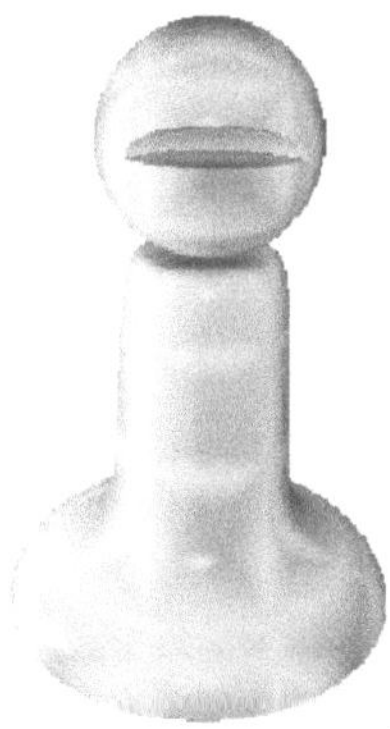

```
<defObject name="pawn">
    <intersectionRf>
        <unionRf>
            <blendedUnionRf a0="0.2" a1="0.3" a2="0.3">
                <aEllipsoid>
                    <vec3 x1="0" x2="0" x3="0"/>
                    <vec3 x1="0.5" x2="0.2" x3="0.5"/>
                </aEllipsoid>
                <blendedDifferenceRf a0="-0.1" a1="0.3" a2="0.2">
                    <aTube>
                        <vec4 x1="0" x2="0" x3="0" x4="0.25"/>
                        <vec4 x1="0" x2="1.3" x3="0" x4="0.20"/>
                    </aTube>
                    <aEllipsoid>
                        <vec3 x1="0" x2="1.4" x3="0"/>
                        <vec3 x1="0.3" x2="0.4" x3="0.3"/>
                    </aEllipsoid>
                </blendedDifferenceRf>
            </blendedUnionRf>
            <translation x="0" y="1.25" z="0">
                <getObject name="head"/>
            </translation>
        </unionRf>
        <aPlane>
            <vec3 x1="0" x2="0" x3="0"/>
            <vec3 x1="0" x2="1" x3="0"/>
        </aPlane>
    </intersectionRf>
</defObject>
```

Fig. 2. A demonstration of the XISL language (script) that defines the implicit "pawn" object and the rendered image of the script.

The XISL implicits are defined by means of the functional representation [18]. The general definition of XISL objects provides implementing of various forms of implicits. Each implicit function is represented via an n-ary hierarchical tree, leafs of which stand for arbitrary implicit primitives and inner nodes stand for deforming and affine transformations as well as for set-theoretic, blending, and interpolation operations.

Several similar modeling systems based on implicits were developed [1, 27]; nevertheless, the other implicit modeling systems were found unsuitable mainly due to the Windows or Linux incompatibility. The main reason to support both operating systems resides in the grid-based implementation [11]. Another important factor, which influenced our decision, was related to function evaluation slowdown in the case of complex Hyperfun models. We performed a comparison of the XISL and Hyperfun systems, based on direct ray tracing of scenes with increasing complexity [10]. Here, XISL was found to be faster in the case of ray tracing of scenes composed of large number of objects. We did not test the BlobTree system.

Cell Geometry

The initial step in muscle cell modeling involves creation of the central skeleton of the cell, which is represented by a system of parallel cross-sectional graphs (c-graphs) distributed along the longitudinal axis. The longitudinal axis is often defined by the orientation of myofibrils. The transversal cell direction is perpendicular to the longitudinal cell axis. The important modeling feature is that each organelle has a preferred orientation along one of these two directions (Fig. 3).

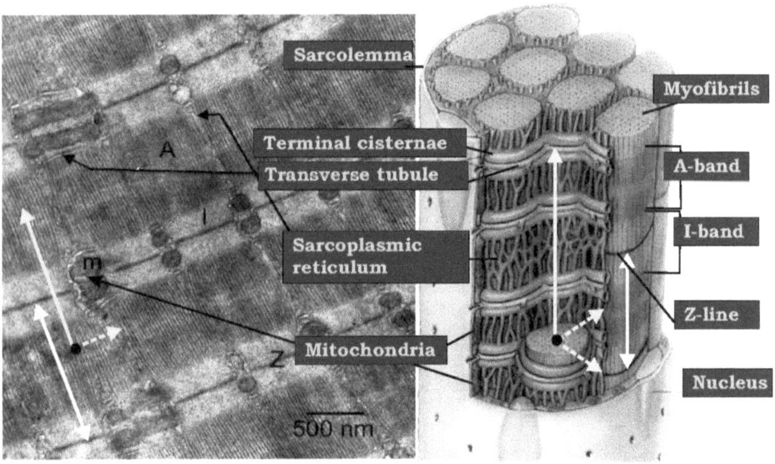

Fig. 3. An electron microscopic image of a longitudinal section of a skeletal muscle cell with a superimposed artistic painting of a muscle cell segment. The yellow arrow depicts the longitudinal cell direction, the dashed yellow arrow represents the transversal cell direction, and the white double arrow delimits the basic repetitive unit—the sarcomere (Courtesy of Dr. Novotova).

The C-graph

C-graphs represent the basic modeling units for cell creation, i.e., produce underlying skeletons required for modeling of all desired organelles. We define a c-graph as a continuous planar graph which defines a finite number of closed non-intersecting polygons. A c-graph is continuous, if each vertex has at least two adjacent edges. An example of such a c-graph is depicted in Fig. 4a.

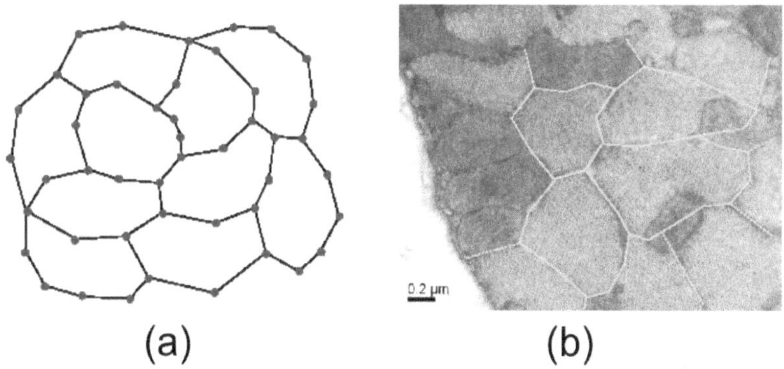

Fig. 4. (**a**) A c-graph defining 10 closed polygons. (**b**) An electron microscopy image of a transversal section of a muscle cell. Edges of the c-graph represent boundaries of neighboring myofibrils (Courtesy of Dr. Novotova).

The c-graph represents boundary interface of neighboring myofibrils, where edges lie near the boundary of myofibrils. Thus, each myofibril is bounded inside a single polygon. This graph is prepared according to a real electron microscopic image of a transversal cell section (Fig. 4b). Such a prepared c-graph is then stored in the c-graph database containing additional information about the number and the area of participating polygons and the c-graph circumference. A c-graph approach is directly exploited to create the myofibrillar system by means of the F-rep of polygons [17] and quadratic interpolation (Fig. 5). The remaining organelles (sarcoplasmic reticulum, mitochondria, t-tubulus and sarcolemma) are defined by means of underlying skeletons derived from the c-graphs.

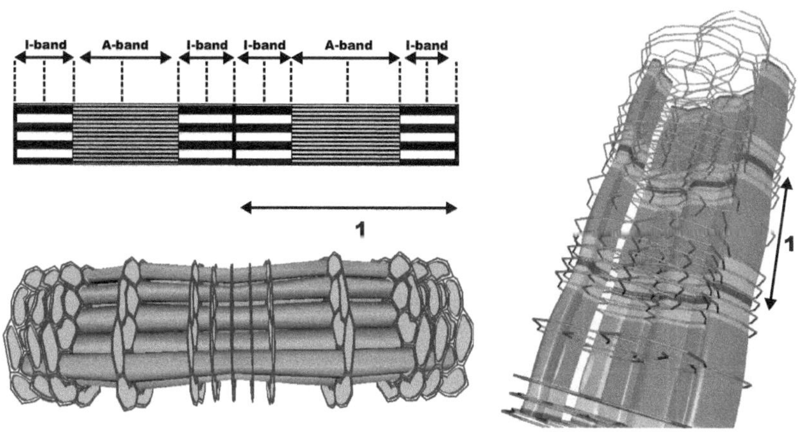

Fig. 5. *Left*: An example of c-graph distribution on two sarcomeres (1—a single sarcomere). Dashed vertical lines at the top represent the distribution of c-graphs. *Right*: Demonstration of the myofibrils and the c-graphs within three sarcomeres; for better clarity the myofibrils located in the middle of the cell are hidden.

Extended Interpolation

The most complex organelle modeled by our tools is the sarcoplasmic reticulum. The extended interpolation technique, which was originally proposed to create its complex network of interconnected cisterns and tubes was later generalized to a complete free-form modeling environment [14].

Sarcoplasmic reticulum (SR) is a membranous structure consisting of two compartments. The first, terminal cisterns of the SR, closely follows the c-graphs in the transversal direction. The second, longitudinal SR, is represented by a system of tiny randomly interconnected and longitudinally oriented tubules, creating a typical mesh (see Fig. 3).

Sarcoplasmic reticulum is modeled in two steps. The first step is to build the system of the longitudinal tubules (F_{lt}), which is achieved by extended interpolation [14, 20] between sets of circular implicit shapes (Fig. 6a). The second step comprises modeling terminal cisterns (F_{tc}), defined in transversal planes, that form a smooth junction to the system of longitudinal tubules. The resultant object is obtained by CSG union of these two components $F_{sr} = F_{lt} \cup F_{tc}$ (Fig. 6b) [14, 5].

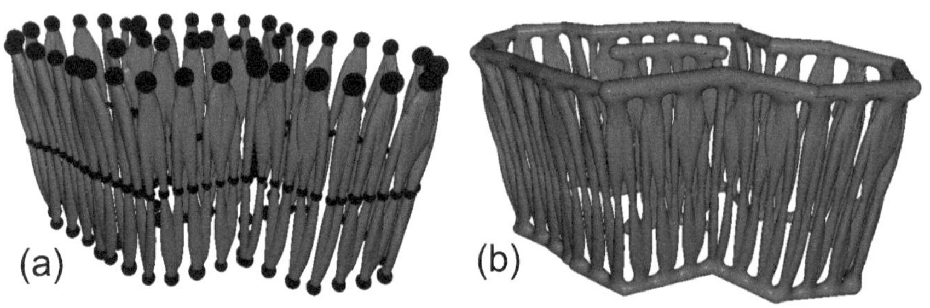

Fig. 6. Modeling of sarcoplasmic reticulum. (**a**) The extended interpolation provides smooth interconnections between circular shapes, here depicted as red spheres. (**b**) The required smooth junction between terminal cisterns and SR tubes is obtained by the blended union.

Representation of Mitochondrial Shape

Mitochondria are closed, membrane-bound, elliptically shaped organelles of irregular smooth forms and variable sizes. In our earlier studies [15], they were represented as convolution surfaces created on underlying line segments. This approach has several drawbacks; the shape of convolution surfaces is axially symmetric, which is not suitable in modeling of the prolonged and deformed mitochondria. Implementation of just linear segments can be thought of as a disadvantage as well. This can be, however, circumvented by using a parametric curve as the convolution primitive; nevertheless, the necessary enumeration of the required convolution integral is in this case very costly.

Implicit Sweep Objects for Mitochondria Modeling

In order to capture the varying elliptical shape of mitochondria, we developed a new method based on implicit sweep objects. The basic components of sweep objects are a 2D sweep template and a 3D sweep trajectory. Several works addressed the problem of creation of generalized sweep objects [17, 21, 23]. The advantage of implicit representation is that curve self-intersection problems are solved thanks to unambiguous point classification. In general, to create a sweep object, a transformation is created that maps a 2D template point **c** to a 3D point **p**. We assume that the transformation maps the center of the 2D template to the point that lies on the trajectory.

In the case of mitochondrial modeling, to preserve the cross-sectional elliptical shape of mitochondria, the 2D template is defined as a 2D implicit ellipsoid with variable dimensions

$$f_e(\boldsymbol{c}) = 1 - \left(\frac{c_x}{r_x(t)}\right)^2 - \left(\frac{c_y}{r_y(t)}\right)^2, \tag{1}$$

where $\mathbf{c} = (c_x, c_y)$, and $r_x(t)$ and $r_y(t)$ define the spatially varying elliptical shape. In general, the trajectory can be represented as an arbitrary parametric curve; however, to define the transformation, it is necessary to find for an arbitrary 3D point $\mathbf{x}$ its nearest point lying on this trajectory. If $C(t)$ describes the trajectory, all its nearest points $NP(\mathbf{x})$ to a given 3D point $\mathbf{x}$ (Fig. 7a) are defined using the following prerequisite:

$$NP(\boldsymbol{x}) = \{C(t) : (\mathbf{x} - C(t)).C'(t) = 0\}. \tag{2}$$

For parametric curves that are of degree 3 and higher, the resultant polynomial of (2) is at least of degree 5. The analytical solution is then unattainable; therefore, we adopt the uniform quadratic B-splines to represent the sweep trajectory:

$$C(t) = [t^2, t, 1]\frac{1}{2}\begin{bmatrix} 1 & -2 & 1 \\ -2 & 2 & 0 \\ 1 & 0 & 0 \end{bmatrix}\begin{bmatrix} \mathbf{P_i} \\ \mathbf{P_{i+1}} \\ \mathbf{P_{i+2}} \end{bmatrix}. \tag{3}$$

Now, the resultant polynomial (2) is of degree 3 and the nearest curve points can be obtained analytically [10].

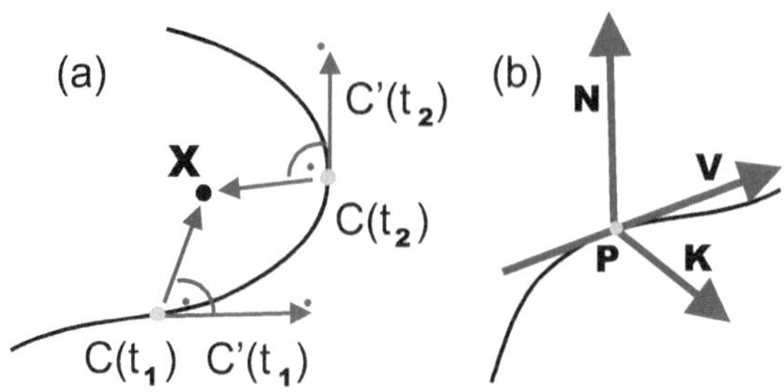

Fig. 7. (**a**) Computing the nearest points of $C(t)$ to the point **x**. (**b**) The mitochondrial reference frame, where $\mathbf{P} = C(t)$, $\mathbf{V} = C'(t)$, $\mathbf{K}\cdot\mathbf{V} = 0$, and $\mathbf{N} = \mathbf{K}\times\mathbf{V}$.

Template Mapping

At each point $\mathbf{P}\in NP(\boldsymbol{x})$ we define a mitochondrial reference frame (MRF) as depicted in Fig. 7b. The transformation that maps the 2D template to the MRF is then defined by an affine transformation matrix:

$$\mathbf{x} = (c_x, c_y, 1)\begin{bmatrix} K_x & K_y & K_z \\ N_x & N_y & N_z \\ P_x & P_y & P_z \end{bmatrix}, \tag{4}$$

which, if computed for a point $\mathbf{P}=C(t)$, is denoted as $\mathbf{M(P)}$. Moreover, in implicit representation it is necessary to compute its inverse matrix $\mathbf{M}^{-1}\mathbf{(P)}$ (Fig. 8a). Note that the elliptical cross-shape is now aligned to two main axes, which is a limiting feature and, thus, also limits direct usage of this approach to model the varying shapes of mitochondria. Sufficiently greater shape variability can be achieved by rotating the 2D sweep template around the trajectory. To yaw the elliptical shape around this curve, we define an angle function $\alpha=\Phi(t)$ that represents rotation of the MRF around the curve tangent **V** (Fig. 8b). The rotation itself, using quaternions, is defined as follows. Considering quaternions $\hat{q} = (N_x, N_y, N_z, 0)$ and $\hat{r} = \left(V_x \sin(\alpha/2), V_y \sin(\alpha/2), V_z \sin(\alpha/2), \cos(\alpha/2)\right)$ the resulting quaternion $\hat{r}\hat{q}\hat{r}^{-1}$ rotates **N** around **V** by the angle α. The same approach is used for the vector **K**.

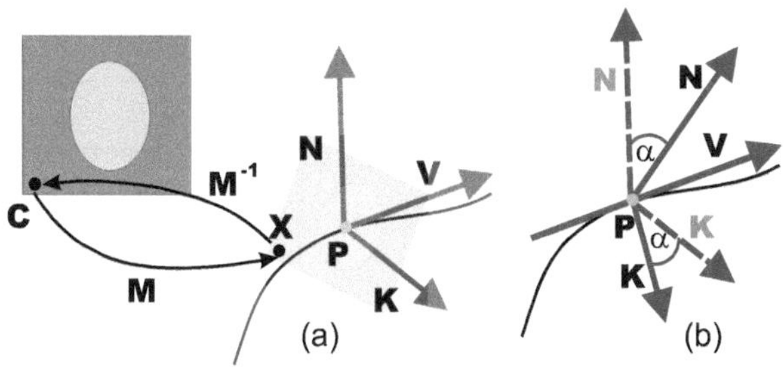

Fig. 8. (**a**) Transformation matrix **M** maps the template to the MRF. (**b**) Rotation of the MRF around **V** by α.

Endcaps

The last task that has not been discussed yet is creating the endcaps. In general modeling sweeps are capped by interpolating the field function to zero. In [21] the authors used another popular form of capping by a CSG difference operation with a half-space. However, the mitochondrial shape is rounded everywhere and, therefore, we define the caps as implicit semi-ellipsoids f_0 and f_1 on both ends with radii $r_z(0)$ and $r_z(1)$, which correspond to the $r_x(k)$, $r_y(k)$ for k=0,1 from (1) (Fig. 9, left). Finally, the resultant formula for a single mitochondrion is given by

$$F_{mt}(\mathbf{x}) = \bigcup_{P \in NP(x)} f_e\left(\mathbf{x}.\mathbf{M}(\mathbf{P})^{-1}\right) \cup f_0\left(\left(\mathbf{x} - C(0)\right).\mathbf{N}_0^{-1}\right) \cup f_1\left(\left(\mathbf{x} - C(1)\right).\mathbf{N}_1^{-1}\right), \tag{5}$$

where both ellipsoid rotation matrices are

$$\mathbf{N}_0 = [\mathbf{K} \quad \mathbf{N} \quad C'(0)]^T \text{ and } \mathbf{N}_1 = [\mathbf{K} \quad \mathbf{N} \quad C'(1)]^T. \tag{6}$$

Figure 9 demonstrates such a mitochondrion defined by a trajectory specified by means of control points.

Test Data

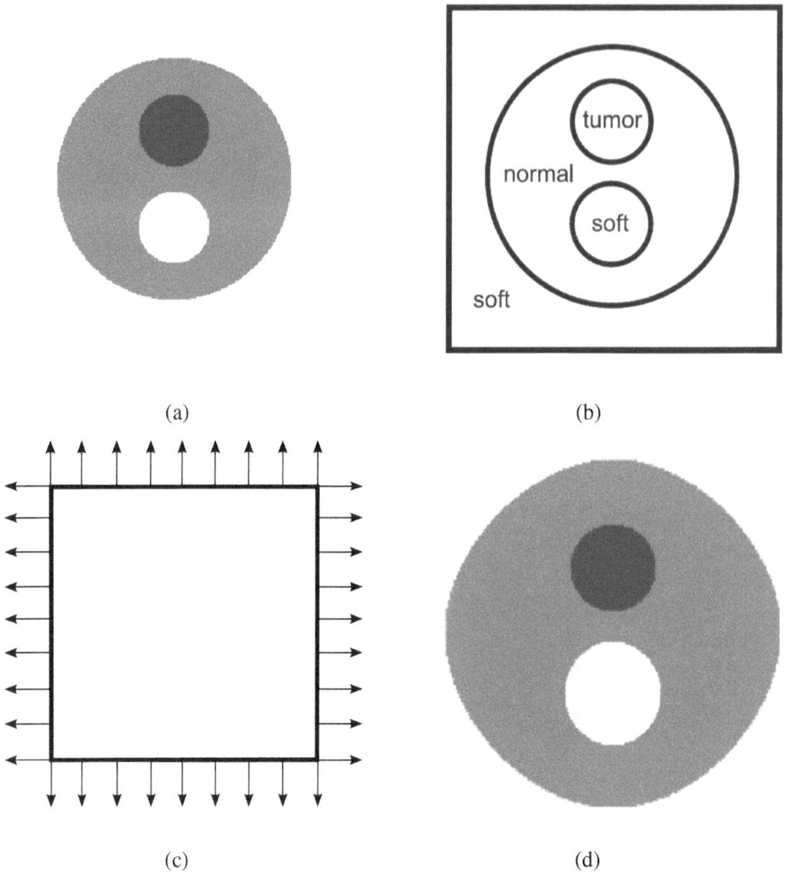

Fig. 1. (**a**) Input image, (**b**) segmentation of the input image, (**c**) representation of boundary condition image with outward expansion, and (**d**) deformed input image.

A set of synthetic data are used to validate the method. These data are generated so it resembles an MRI image of a brain, as seen in Fig. 1a. To simplify notations three tissue types are considered. Namely, normal matter (gray color), tumoral tissue (black color), and soft tissue (white color), depicted in Fig. 1b. Second, a boundary condition image is constructed as seen in Fig. 1c, where the displacement vectors are shown. In the following, the boundary condition image is described by $\mathbf{y}$. In this example, the boundary condition image describes an outward expansion. In real applications, the boundary condition image can, for example, be computed using a non-rigid registration algorithm [3, 16]. To be able to control the amount of confidence in the boundary condition image a confidence of models, which could lead to several hours of computation on a single CPU. Therefore, parallelization of this process is necessary.

We implemented the GeomCell modeling system in a grid environment, which besides parallelization of the computation itself enables to implement additional tools fulfilling user demands (e.g., visualization). A user can specify his/her requirements interactively through a web portal (Fig. 10), which additionally makes possible to observe the state of active computational jobs and results in the form of images and computed VSDs.

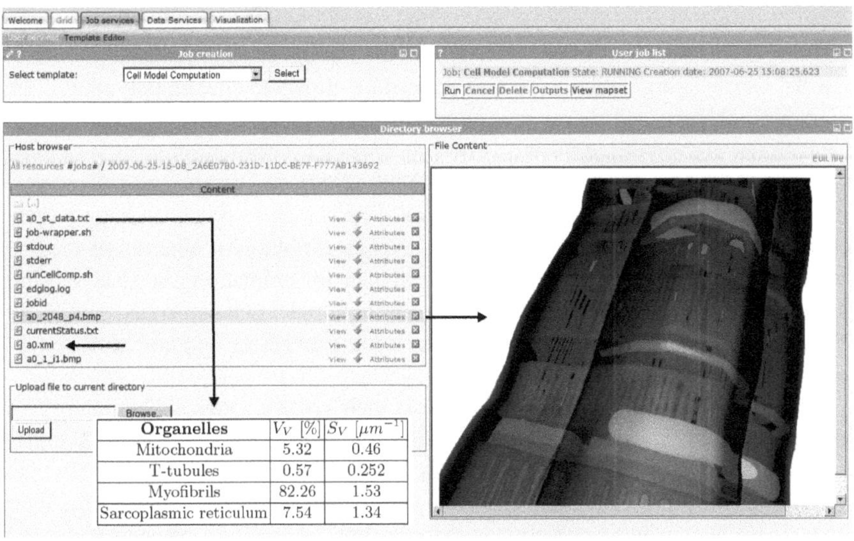

Organelles	V_V [%]	S_V [μm^{-1}]
Mitochondria	5.32	0.46
T-tubules	0.57	0.252
Myofibrils	82.26	1.53
Sarcoplasmic reticulum	7.54	1.34

Fig. 10. A screenshot demonstrating the GeomCell portal GUI that enables users to prepare, submit, and verify their models. The `a0_st_data.txt` represents the file that contains the evaluated volume and surface densities for each organelle class (the bent arrow), the file `a0.xml` includes XISL definition of the organelles (*left arrow*), and `a0_2048_p4.bmp` is one of the rendered images per `a0` model (*right arrow*). The remaining files, visible in the list on the left side, were generated automatically.

From the engineering point of view, it was necessary to integrate several technologies into a distributed system capable of providing computational power that supports heterogeneous computing platforms, means for collaborative result sharing of eScience experiments, and provides easy to use user interface.

The use of grid computing technology provided us with speedup of stereological verification results delivery to the end user. The speed up was linear, depending only on the number of worker nodes involved; only relatively small grid middleware overhead was observed. It is important to mention that the development and tests were performed on lightly loaded infrastructure and thus the wait time of submitted job executables in the queues of batch systems was minimal. The tests of the system were performed on the GILDA [9] testbed—test and training infrastructure of EGEE project [6]. The implementation of the cell tools in a grid environment can be found in [11, 12].

Future Work

The presented model is currently only a "static" one. In order to incorporate behavior of organelles and cells, we will have to add a physical layer to all objects. There are several important topics that need to be solved: collision detection resulting in deformations of organelles, realistic muscle contraction, and growth of objects. Our existing collision-solving procedure allows to change only one of the interacting objects, and therefore a more advanced method will be required. For instance the algorithm based on [8], where a special deformation term is defined and added to both implicit functions of colliding objects, can be used. However, first it is necessary to specify the properties of deformable objects and apply them to the skeleton upon which the implicit layer is created. An alternative method could be based on deformable surfaces designed for animation called active implicit surfaces [3]. This approach allows any topological changes, and similar to snakes in computer vision, the surface can track a target and exhibit surface tension under its own criteria.

An important behavior of living objects is their growth. Growing human organs can be described as dynamical systems with a dynamical structure [4]. In such systems not only the values of variables characterizing system components but also the number of components and the connections between them may change over time. There is a need to construct a mathematical description of such a system. The same principles apply to the muscle cells. The simulation of human organ growth can be seen as an imitation of the reality for studying the effect of changing parameters in a model as the means of preparing a decision or predicting experiment results.

In the implementation of physical layer, however, numerous challenging tasks can appear, e.g., connection between growing simulations and realistic deformations, collisions of multiple objects.

Conclusion

In this chapter we proposed a methodology for construction of muscle cells models and cell organelles. The organelle representation is based on the theory of implicit surfaces. We used the XISL package as the supporting modeling system for implicits. The modeled organelles include myofibrils, mitochondria, t-tubules, sarcoplasmatic reticulum, and sarcolemma. Specifically, here we focused on design of mitochondrial shape and geometry. Natural variability of size, shape, and position of organelles is achieved by random variation of the specific parameters within given limits in MDL configurations. The resulting model is materialized by evaluation of implicit functions and is adjusted for a given set of specific parameters defined by a biologist. The whole system, called GeomCell, was implemented within the grid environment.

Acknowledgments We would like to thank Alexandra Zahradníková (Institute of Molecular Physiology and Genetics, SAS) for consultation and helpful comments regarding this chapter. This work was supported by the Slovak Research and Development Agency under Contract No.APVV-20-056105 and Contract APVT-51-31104 and by the grant VEGA No. 2/0174/09.

References

1. Adzhiev V, Cartwright R, Fausett E, Ossipov A, Pasko A, Savchenko V (1999) Hyperfun project: A framework for collaborative multidimensional f-rep modeling. In: Proc. of the Implicit Surfaces '99 EUROGRAPHICS/ACM SIGGRAPH Workshop, pp. 59–69
2. Bloomenthal J, Bajaj C, Blinn J, Cani-Gascuel MP, Rockwood A, Wyvill B, Wyvill G (1997) Introduction to implicit surfaces. Morgan Kaufman Publisher Inc, San Francisco, CA
3. Desbrun M, Gascual MP (1998) Active implicit surfaces for animation. In: Proc. Graphics Interface, pp. 143–150
4. Durikovic R, Kaneda K, Yamashita H (1998) Animation of biological organ growth based on l-systems. Computer Graphics Forum 17(3):1–14
5. Ďurikovič R, Czanner S, Parulek J, Šrámek M (2008) Heterogeneous modeling of biological organs and organ growth, In the book on "Heterogeneous Objects Modeling and Applications", Lecture Notes in Computer Science, A Pasko, V Adzhiev, P Comninos (Eds.), Springer-Verlag, Berlin, Heidelberg, Vol. 4889, pp. 239–258
6. EGEE: Enabling grids for e-science (2008). http://www.euegee.org/
7. Felkel P, Fuhrmann A, Kanitsar A, Wegenkittl R (2002) Surface reconstruction of the branching vessels for augmented reality aided surgery. In: Biosignal 2002 Vol. 16, pp. 252–254
8. Gascuel MP (1993) An implicit formulation for precise contact modeling between flexible solids. In: SIGGRAPH '93: Proceedings of the 20th Annual Conference on Computer Graphics and Interactive Techniques, ACM Press, pp. 313–320
9. GILDA: Gilda testbed home page, https://gilda.ct.infn.it/
10. Parulek J (2007) Problem solving environment for stereology based implicit modeling of muscle cells. PhD thesis, Faculty of Mathematics, Physics and Informatics, Comenius University
11. Parulek J, Ciglan M, Šimo B, Šrámek M, Hluchý L, Zahradník I (2007) Grid problem solving environment for stereology based modeling. In: International Conference on Grid Computing, High-Performance and Distributed Applications, OTM 2007, Part II, Springer-Verlag, Berlin, Heidelberg, pp. 1417–1434
12. Parulek J, Ciglan M, Šimo B, Šrámek M, Hluchý L, Zahradník I (2007) Metadata in problem solving environment for stereology based modeling. In: L. Hluchý et al. 3rd InternationalWorkshop on Grid Computing for Complex Problems, Institute of Informatics, SAS, Bratislava, pp. 126–133
13. Parulek J, Novotný P, Šrámek M (2006) XISL—a development tool for construction of implicit surfaces. In: SCCG '06: Proc. of the 22nd Spring Conference on Computer Graphics, Comenius University, Bratislava, pp. 128–135
14. Parulek J, Šrámek M (2007) Implicit modeling by metamorphosis of 2d shapes. In: SCCG '07: Proceedings of the 23nd Spring Conference on Computer graphics, Comenius University, Bratislava, pp. 227–234

15. Parulek J, Šrámek M, Novotová M, Zahradník I (2006) Computer modeling of muscle cells: Generation of complex muscle cell ultra-structure. Imaging & Microscopy 8(2):58–59
16. Parulek J, Zahradnik I, Sramek M (2004) A modelling tool for construction of 3d structure of muscle cells. In: J. Jan, J. Kozumplik, I. Provaznik (Eds.), Analysis of Biomedical Signals and Image proceedings, Brno University of Technology, Czech Republic, pp. 267–269
17. Pasko A, Savchenko A, Savchenko V (1996) Polygon-to-function conversion for sweeping. In: J. Hart, K. van Overveld (Eds.), The Eurographics/SIGGRAPH Workshop on Implicit Surfaces, Eindhoven, The Netherlands, pp. 163–171
18. Pasko AA, Adzhiev V, Sourin A, Savchenko VV (1995) Function representation in geometric modeling: Concepts, implementation and applications. The Visual Computer 11(8):429–446
19. Pasko AA, Savchenko VV (1994) Blending operations for the functionally based constructive geometry. In: CSG 94 Set-theoretic Solid Modeling: Techniques and Applications, Information Geometers,Winchester, UK, pp. 151–161
20. Savchenko V, Pasko A (1998) Transformation of functionally defined shapes by extended space mappings. The Visual Computer 14(5–6):257–270
21. Schmidt R, Wyvill B (2005) Generalized sweep templates for implicit modeling. In: GRAPHITE '05: Proc. of the 3rd International Conference on Computer Graphics and Interactive Techniques in Australasia and South East Asia, ACM Press, New York, NY, USA, pp. 187 196
22. Smith C, Prusinkiewicz P (2004) Simulation modeling of growing tissues. In: 4th International Workshop on Functional-Structural Plant Models, pp. 365–370
23. Sourin AI, Pasko AA (1996) Function representation for sweeping by a moving solid. IEEE Transactions on Visualization and Computer Graphics 2(1):11–18
24. Tait R, Schaefer G, Kühnapfel UC¸ Akmak H (2003) Interactive spline modelling of human organs for surgical simulators. In: 17th European Simulation Multiconference, pp. 355–359
25. Tsingos N, Bittar E, Gascuel MP (1995) Implicit surfaces for semi-automatic medical organ reconstruction. In: RA Earnshaw, JA Vince (Eds.), Computer Graphics International, Leeds, Royaume-Uni, Academic Press, pp. 3–15
26. Weibel E, Bolender R (1973) Stereological techniques for electron microscopic morphometry. In: V.N.R.C. MA Hayat (Eds.) Principles and techniques of electron microscopy, New York, Vol. 3, pp. 237–296
27. Wyvill B, Guy A, Galin E (1999) Extending the csg tree (warping, blending and boolean operations in an implicit surface modeling system). Computer Graphics Forum 18(2):149–158

Tracking Organs Composed of One or Multiple Regions Using Geodesic Active Region Models

A. Martínez[1] and J. J. Jiménez[2]

Abstract In radiotherapy treatment it is very important to find out the target organs on the medical image sequence in order to determine and apply the proper dose. The techniques to achieve this goal can be classified into extrinsic and intrinsic. Intrinsic techniques only use image processing with medical images associated to the radiotherapy treatment, as we deal in this chapter. To accurately perform this organ tracking it is necessary to find out segmentation and tracking models that were able to be applied to several image modalities involved on a radiotherapy session (CT, MRI, etc.). The movements of the organs are mainly affected by two factors: breathing and involuntary movements associated with the internal organs or patient positioning. Among the several alternatives to track the organs of interest, a model based on geodesic active regions is proposed. This model has been tested over CT images from the pelvic, cardiac, and thoracic area. A new model for the segmentation of organs composed by more than one region is proposed.

Introduction

In a radiotherapy process, an important task consists of localizing the organ of interest in order to be properly radiated. Finding out the borders or contours of the structure of interest is important too. If we know the contours, the damage received by the nearest tissues to the target organ may be reduced. Taking medical images (CT, MRI, EPI, etc.) is a common stage during a radiotherapy treatment.

Those images, from the patient's body, are taken in order to control the position and dose. The application of the tracking model to these images makes possible to set up the radiation direction and dose [25].

The patient organs are mainly affected by anatomical movements, breathing is the main one, but there are other patient movements (voluntaries and involuntaries)

[1] Departamento de Informática, University of Jaén, Jaén, Spain
amalbala@ujaen.es
[2] Departamento de Informática, University of Jaén, Jaén, Spain
juanjo@ujaen.es

N. Magnenat-Thalmann et al. (eds.), *Recent Advances in the 3D Physiological Human*,
DOI 10.1007/978-1-84882-565-9_3,

that have to be considered too. The problems associated with these movements may be classified using a criterion based on the area affected by the radiation:

- At the periphery of the target volume: trying to determine the organ contour.
- Within the target volume: studying the effect of the organ movement because the radiation dose may be heterogeneous.
- Just outside the target volume: what organs are at risk to be radiated.

There are interesting studies about the movements associated with the organs, like breathing and involuntary movements [25, 34], or specific studies when these movement are affected by a radiation session [23, 7, 35]. Breathing is considered the main cause of organ movements [4, 9, 11, 35, 36, 40, 42].

If we try to control the movement caused by breathing, two alternatives may be considered:

- To control the patient behavior [19]
- To apply methods for tracking the organs

To study the first alternative, to control the patient behavior, some methods about training the patient for a proper breathing control, like deep inspiration breath hold (DIBH) [19] are available. Taking into consideration the second alternative, some techniques for controlling and tracking the organ movement can be used. A classification of these techniques could be

- extrinsic, if external objects are introduced inside the image space (markers, adaptors) [26];
- intrinsic, if they are based only on image features, natural markers (bones, anatomical markers), segmentation methods, and pixel/voxel intensity properties.

There are several techniques and methods that can be applied to tracking structures or organs. Some of them are point distribution model (PDM) or Snakes [1, 2, 12, 13, 17, 31, 22, 37], Kalman filter [8, 39], segmentation methods based on gradient and morphological operation like erosion and dilatation [32], geometrical models [24, 29, 28] based on camera calibration [10], applying register [18, 20, 16], using commercial specialized system as CyberKnife [9, 36], or using external and internal markers [5, 41, 36]. A deep review about tracking organs in real time is given by Murphy [25]; a study about movements associated to breathing and how to manage them is described by Keall and Magerast [15] and about tracking radiotherapy environments by Shirato et al. [35].

Although the main movement that affects the organs is breathing, other movements must also be taken into consideration as filling of internal organs (i.e., bladder), cardiac movements. We propose a model to track several organs only with the information from image features without using any artifacts (markers). This model is based on geodesic active regions (GAR) [29, 30, 28] with some extensions over the classical model. A common problem with anatomical images is that the target organ is defined by multiple regions. This issue could appear if the organ was partially filled with a contrast fluid (contrasted CT) or if the organ is composed of tissue and air (rectum). A new model to manage organs that are composed of multiple regions is proposed.

Image sequences taken on the radiotherapy process usually have small changes on size, shape, and position between frames. If we use this characteristic it is not necessary to apply a complete segmentation process between frames. Reusing previous results is possible by setting the final contour of the previous frame as the initial contour for the next one.

The tracking extension from classical geodesic active region is explained in the next section, and the new model to track organs composed of multiple regions is explained too. Some tests with both models applying to images from thoracic and pelvic area are described in the third section, and a validation of this model compared with medical delineation is show in the fourth section. Finally on the last section the conclusions are presented.

Methods and Materials

In this work the organ is tracked using GAR [29, 30, 28] applied to medical image. We extend the classical model to deal with medical images, adapting the equation from the classical approach to the features of the image modality and the target organ. Studying the problem of organs composed of several regions, a new model based on the previous one has been developed. Both models have been tested over CT sequences from pelvic and thoracic area, the target organs being bladder, heart, and lungs, respectively. These organs have been selected because they are the most frequent ones involved in a radiotherapy process.

The active region model has good efficiency on image segmentation independent of the initial position, opposite to the classical geometric models, *level set* [21, 33, 27], and the geodesic models [6], against the classical snake or PDM model [14], the model proposed has more computational cost but it is more accurate, can easily handle topological changes, and is independent of the initial contour (control points). These classical models are influenced by the initial contour and they force it to be totally inside or outside the region of interest (Fig. 1). So the GAR model has the advantages of geodesics models like handling topological changes and obtaining good results with noisy images, easy application to 3D image, independent of the initial contour, and finally, the combination of contour and region features make the model robust against illumination changes.

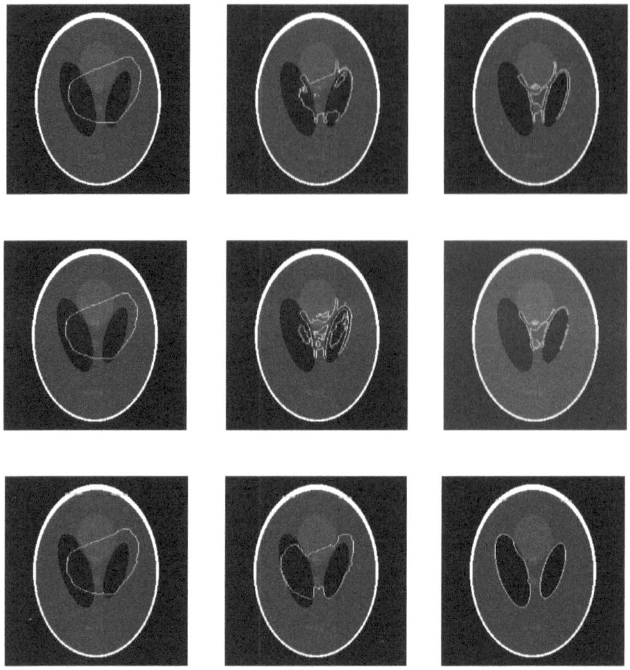

Fig. 1. Results of applying the geometric model (first row), the geodesic model (second row), and the geodesic active region model (third row). If the initial contour is not totally inside or outside the region of interest the geodesic and geometric models have wrong results, but the GAR model overcomes this problem.

Now the extension of the geodesic active region model, in order to apply it to medical images, and the new model to manage organ composed of multiple regions are explained.

Geodesic Active Region Model

This model was initially developed for supervising texture segmentation [29] and then to movement estimation [30] and tracking. The initial assumption is of only one target object. The geodesic active region model is composed of two modules associated with the main characteristics on tracking and segmentation. First the contour module tries to attract the curve to the target region border, which is based on intensity changes on the contours (gradient). Second the region module tries to segment the region carrying out some statistical and homogeneity properties that are associated with the target region.

Contour Module

The geodesic model [6] is the segmentation technique used in this contour model. The model is a combination of the geometric model [21, 33, 27] based on the heat equation and the point distribution model (PDM) [14]. The equation that defines the geodesic model is

$$E(v(s)) = \int_0^1 g(|\nabla I(v(s))|^2)ds + \int_0^1 |\dot{v}(s)|^2 \, ds \tag{1}$$

Where $v(s)$ is the curve or contour that defines the geodesic, g is a monotonic decreased function $g(r) \rightarrow 0$ if $r \rightarrow \infty$ (in the geodesic model g is a stop criterion [6, 27]), and $\nabla I(v(s))$ is the image gradient. Taking into consideration a regularization term, the previous equation may be defined using probabilistic definitions [29]. So $v(s)$ is now ∂R, and the previous function to localize the contour $|\nabla I(v(s))|^2$ is now expressed using probabilistic functions p_C so the model equation (2) is defined as follows:

$$E(\partial R) = \underbrace{\int_0^1 g\left(\underbrace{p_C(I/\partial R(c))}_{Contour\ Probability}\right)}_{ContourAttraction} \underbrace{\left|\partial \dot{R}(c)\right|}_{Regularity} dc \tag{2}$$

where ∂R is a contour parameterization.

Region Module

Now we need to define the region that has the desired features. Assuming that the region of interest is A and p_A is the probabilistic function that defines the distance from the actual region to the ideal one, B is the rest of the image and p_B is the probabilistic function so similar to the previous one. Finding the optimal partition is similar to obtaining the partition that maximizes the probability p_A and minimizes p_B. Probability maximization is similar to the minimization of (–log) function probability (and converts the product on probabilities sum) [3]:

$$E(\partial P(R)) = -\iint_{R_A} \log\left[\underbrace{p_A(I(x,y))}_{A\ probability}\right] dxdy - \iint_{R_B} \log\left[\underbrace{p_B(I(x,y))}_{B\ probability}\right] dxdy \tag{3}$$

Each component is defined using the regions A and B defined by the curve, and it tries to maximize the p_A probability from the image I. The image regions are separated applying the properties associated with the intensities. In order to extend the classical GAR model to apply it to medical images, a statistical approach to define the probability functions has been used. These functions are calculated from the image histogram. The extended model studies the histogram of the region inside the contour, obtaining standard deviation and intensity average. Using these measures a distance function to each image pixel and a probability distribution function based on each pixel belonging to the target regions is defined too. This function is based on quadratic error. Applying the quadratic error to determine the distance from the pixel intensity to the statistical region properties, the equation that defines the distribution associated with the region is (4)

$$\text{Region Module} = \sqrt{[(\overline{u} \pm \sigma) - i]^2} \tag{4}$$

where $\overline{u}$ is the region average, σ is the standard deviation, and i is the pixel intensity. There are several functions and statistical measures that may be defined based on the target and the image modality. In this work, we have used two region functions, one based on the most frequency intensity $\hat{u}$ and other based on the intensity average $\overline{u}$. If the image has well-defined tissues with well-defined intensities (lung or pelvic images) then the most frequency intensive function is used, but if the images have noise the intensity average function works better (cardiac images).

Geodesic Active Region General Function

Combining the contour module (2) and the region module (3), the geodesic active region equation is obtained:

$$E(\partial P(R)) = (1-\alpha)\underbrace{\int_0^1 g(p_C(I(\partial R(c))))\,|\,\partial \dot{R}(c)\,|\,dc}_{Contour\ Module} - \sum_{X \in A,B} \alpha \underbrace{\iint_{R_X} \log[p_X(I(x,y))]dxdy}_{\text{Re}\,gion\ Module} \tag{5}$$

where α is a constant that defines the weight of each term in the tracking. Applying the functions previously defined to adapt the classical GAR model to medical images, the modules of general equation are defined as follows:

$$E(\partial P(R)) = \underbrace{(1-\alpha)\int_0^1 g(|\nabla I(R(c))|)dc + \int_0^1 |\partial \dot{R}(c)|\, dc}_{Contour\ Module}$$
$$\underbrace{-\alpha\int_0^1 \sqrt{[(u+\sigma)-i]^2 + [(u-\sigma)-i]^2}}_{Region\ Module} \tag{6}$$

Reusing Partial Results

The tracking method has been defined, and this model can be applied to each frame separately, but a better approach consists of reusing previous results. The changes on size, shape, and position between frames are small and usually they occur simultaneously in each transition. The association between this changes and the breathing movement is clear, because the breathing is an intermittent movement. In the first phase the organ gets bigger and moves in one direction and in the next phase it gets smaller and moves in the opposite direction. On a 3D or volume image the changes between slices are small and similar to the time sequences. So we can use the final contour of the previous frame as the initial contour on the next one. This approach reduces the computational time and makes the initial contour definition an automatic task.

Geodesic Active Multiple Regions

To use contrast to stress the organ in a radiotherapy medical image is a very common technique; the organ filled with the contrast appears brighter than the surroundings tissues. But one problem occurs if the contrast substance does not completely fill the organ. This problem was found on bladder contrast images. In order to stress the bladder with the contrast the organ stands partially filled so only a part of the complete organ is shown. Organs composed of multiple tissues or of air and tissues (rectum) have a similar problem. Both cases are not solved by tracking models so a new model must be defined.

In order to overcome this problem an extension of the basic model has been developed in order to give the model the capability of segmenting an organ composed of more than one region. So in this case the organ of interest is a union of regions, this new model is called geodesic active multiple regions (GAMR).

This approach can be applied to real organs that are composed of different tissues that generate different intensity values on medical images.

Histogram Study

To solve the previous problem, when one organ is composed of more than one region, a study of the characteristics of the organ of interest is performed. In our case the study is focused on the bladder.

In this work a set of 256 bladders have been used. These images are classified into three classes: without contrast, with contrast full filled, and with contrast partial filled. The bounding box of the region of interest (bladder) has been extracted (Fig. 2) and the histogram of the bounding box was studied (Fig. 3).

Analyzing the bladder region histograms and calculating the local highest values, a criterion to find out the values that belong to the region that comprise the organ has been extracted. The histogram maximum is studied and two cases considered:

- If there are two local maximums, the brighter one belongs to the target region and the other one to its background. This model is similar to the GAR applied to medical images previously defined (see first and second histograms in Fig. 3).
- If there are three or more maximums, the lowest value is considered as background and the other values the intensities of the regions of interest (see third histogram in Fig. 3).

Then this maximum values are introduced inside the equation as the region of interest values.

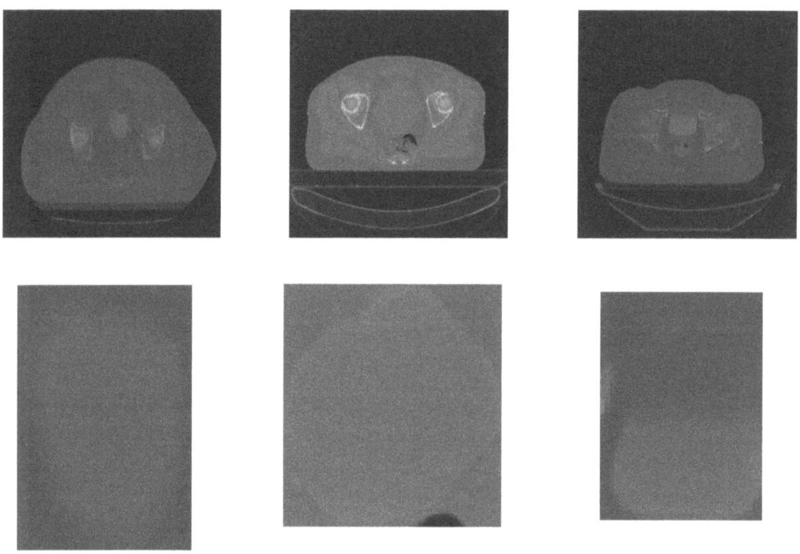

Fig. 2. Three cases: organ full filled (first row), no filled (second row), and partial filled (third row).

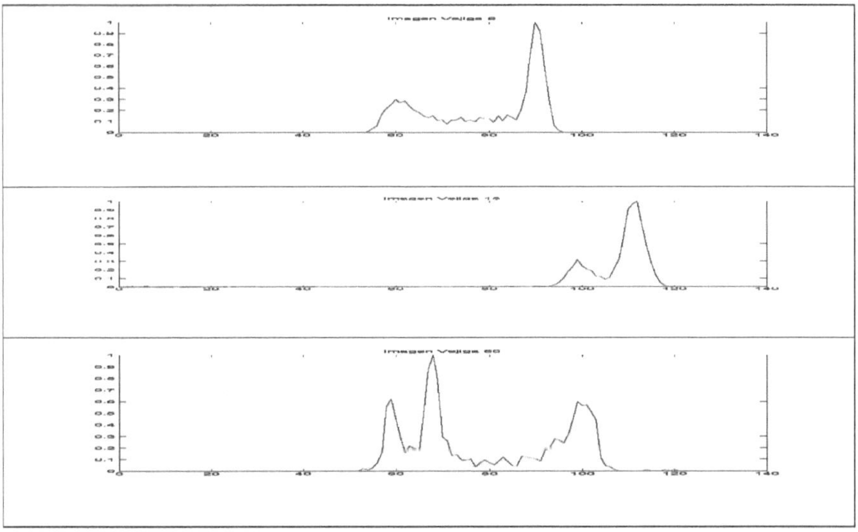

Fig. 3. Histograms associated with the three cases of contrasted organs.

Geodesic Active Multiple Regions Function

In order to insert into the geodesic active region function the concept of more than one region of interest, a linear function of the histogram maximum is applied.

$$p = \frac{d}{m} \tag{7}$$

where p is the final value of the present pixel, d is the distance between the pixel intensity and the value of the interest region, m is the maximum of the region extracted from the study of the histogram, and i is the intensity of the pixel on the original image. With this function the pixels with higher values increase their weight on the function, this is motivated by the assumption that the contrast region is brighter than the other ones (have higher intensity). Applying this equation each pixel has as values number of regions that comprise the organ. These values must be reduced to one in order to apply them on the GAMR equations. The maximum value is selected as the final value of each pixel.

Experimental Results

Some experimental results applying geodesic active region to medical images are presented. In these sequences one target organ is followed over the sequence. First the GAR model has been applied to CT sequences taken from the pelvic area. The target organ is the bladder (Fig. 4). The parameters are α = 0.50, Gaussian with a standard deviation σ = 1.0, and Δt = 0.5. The region module function is based on the most frequency intensity (4), the iterations of each frame and the computational time are shown in the table in Fig. 4.

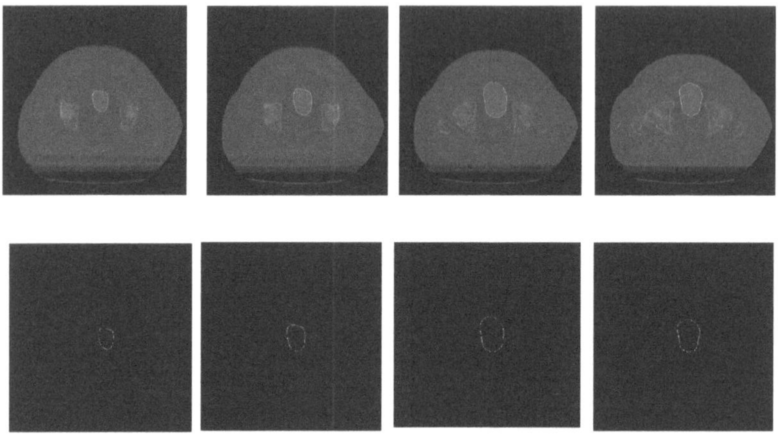

Image	Iterations	Comp. Time (s)
Frame1	10	0,142
Frame2	150	2,627
Frame3	22	0,456
Frame4	21	0,436

Fig. 4. Results applying GAR model to CT sequence from the pelvic area.

Second the GAR model has been applied to follow the heart over an image sequence taken from the thoracic area. In this sequence the heart is affected by breathing and by the intrinsic cardiac movement (Figs. 5 and 6). Different frame intervals have been applied:

- For frames from 1 to 4, frame interval is 1 (Fig. 5).
- For frames from 10 to 20, frame interval is 10 (Fig. 6).

The module region function is based on the region average (4). Finally the model has been applied to tracking the lungs (Fig. 7); in this case the organ of interest is composed of two contours and the model can handle topological changes.

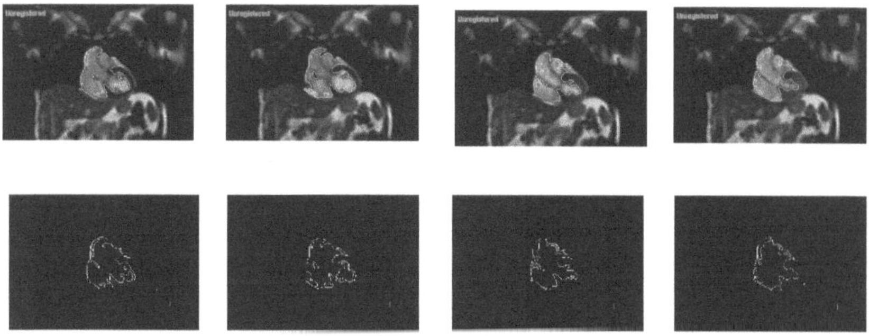

Fig. 5. Results applying GAR model to thoracic area trying to find out the heart. Parameters: $\alpha = 0.25$, Gaussian with a standard deviation $\sigma = 1.0$, and $\Delta t = 0.31$. Frame interval = 1.

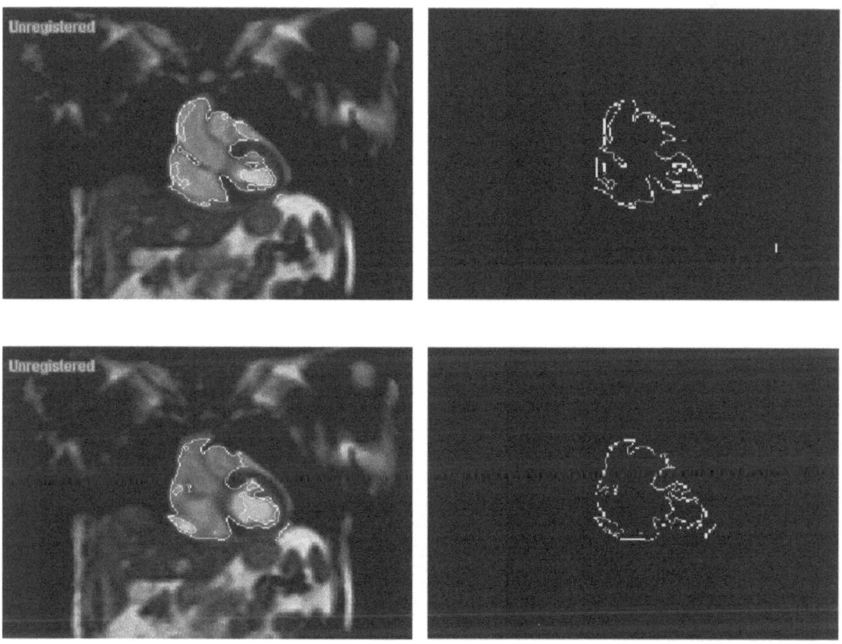

Fig. 6. Results applying GAR model to thoracic area trying to find out the heart. Parameters: $\alpha = 0.25$, Gaussian with standard deviation $\sigma = 1.0$, and $\Delta t = 0.31$. Frame interval = 10.

where **p** is the penalty vector field image. The penalty image is proportional to the confidence image

$$p_{x_s} = Kc_{x_s}, \quad p_{y_s} = Kc_{y_s}, \tag{19}$$

where $\mathbf{c}_s = [c_{xs}\ c_{ys}]^T$, $\mathbf{p}_s = [p_{xs}\ p_{ys}]^T$, and K is the penalty constant. By issuing a penalty, a low energy value is sensitive to the difference between the boundary vector and the displacement vector. Hence, the probability is high when the displacement vector at a site coincides with the boundary condition vector whenever boundary conditions are available, indicated by the confidence vector.

Hierarchical Markov Random Fields

The idea is to propagate the geometrical information through hierarchical levels. Low-resolution scale levels express global geometrical features whereas high-resolution scale levels are used for fine-tuning. Geometrical information is needed to converge to a global solution. By using a hierarchical approach, local energy optimizers can be used to find global solutions. Local optimizers are stable and predictable whereas global optimizers do not guarantee convergence and the computation time is enormous, this is further discussed in the next section.

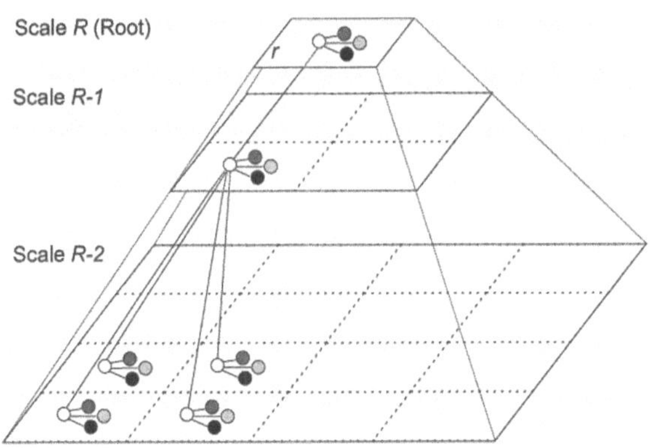

Fig. 3. Dependency graph corresponding to a quadtree structure, with *white*, *light gray*, *dark gray*, and *black circles* representing displacement vectors, boundary condition vectors, confidence vectors, and mechanical property vectors, respectively.

In [14] HMRF for segmentation is presented; it follows the adaptation to the presented method. Let $G = (S,L)$ be a graph composed of a set S of nodes and a set L of edges. A tree is a connected graph that has no cycle. Each node has a unique parent node, except the root node r. A quadtree, as illustrated in Fig. 3, is a special

Validation

A ROC analysis [38] has been used to validate the results from GAR, taking into consideration manual delineation made by the doctors. In this graphics the manual delineation is the red line and the blue one is the GAR segmented line (Figs. 10 and 11).

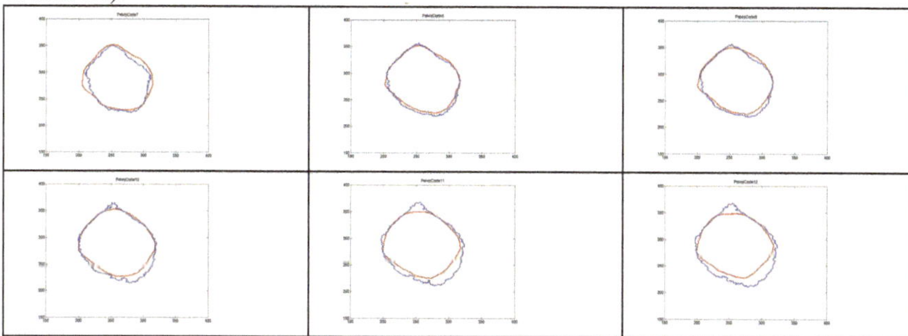

Slice	Sensibility	Specificity
Slice 1	0.8694	0.9999
Slice 2	0.9780	0.9975
Slice 3	0.9601	0.9972
Slice 4	0.9885	0.9940
Slice 5	0.9980	0.9902
Slice 6	0.9960	0.9853

Fig. 10. ROC analysis between manual (*red*) and automatic delineation (*blue*).

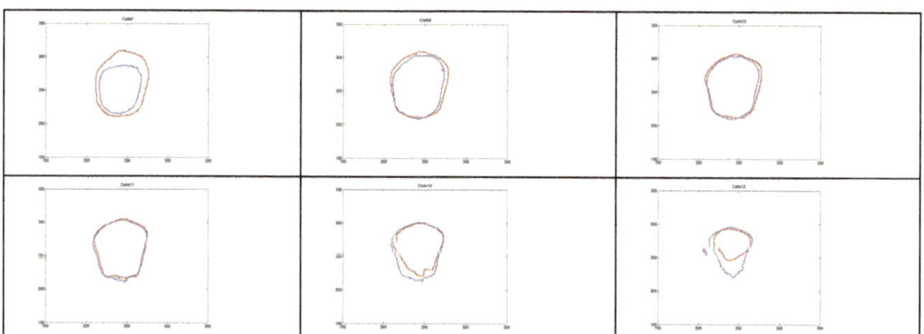

Slice	Sensibility	Specificity
Slice 1	0.8694	0.9999
Slice 2	0.9780	0.9975
Slice 3	0.9601	0.9972
Slice 4	0.9885	0.9940
Slice 5	0.9980	0.9902
Slice 6	0.9960	0.9853

Fig. 11. ROC analysis between manual (*red*) and automatic delineation (*blue*).

Conclusions

A tracking method based on geodesic active regions has been presented. This model extends the classical one in order to be applied as a technique to track organs over medical image sequences. This tracking is a main task on the radiotherapy treatment. Its goal is to find out the target organ and the organs at risk. This information is necessary to determine the radiation dose and radiotherapy sessions duration. The tracking model has been tested over several anatomical images. The results have been quite good in all tests. The target organs are the bladder over a pelvic area and the heart and lungs on thoracic area. The reuse of previous results reduces the computational time and automatizes the initial contour process. Some functions to define the region and contour modules have been presented and applied on the experimental results. The tests have proved that GAR model has quite good results applying it to medical sequences. A ROC analysis has been used to validate the results. The validation compares manual and automatic delineation. The validation study shows high specificity and sensibility values for the automatic delineation using GAR and also better values than manual delineation. A new model, the GAMR, based on organ composed of several regions, is presented and preliminary tests are shown. This model has been applied to organs that are composed of several tissues like rectum or organs partially filled like contrasted bladders. This is a work in progress; at present time we are trying to optimize the model increasing the speed and studying the alternatives to make a real-time system. Predictors and filters are being studied to achieve the real-time goals. More tests applying this model to other image modalities involved in radiotherapy treatment are being studied too.

Acknowledgments This work has been partially supported by the Spanish Ministry of Education and Science and the European Union (via ERDF funds) through the research project TIN2007-67474-C03-03, by the Consejería de Innovación, Ciencia y Empresa of the Junta de Andalucía through the research project P06-TIC-01403, and by the University of Jaén through the research project UJA-08-16-02.

References

1. Akgul Y, Kambhamettu C, Stone M (1999) Automatic extraction and tracking of the tongue contours. IEEE Transactions on Medical Imaging 18:1035–1045
2. Bascle B, Bouthemy P, Deriche R, Meyer F (1994) Suivi de primitives complexes sur une sequence d'Images. Tech. rep., Institut National de Recherche en Informatique et en Automatique (INRIA)
3. Berbeco R, Jiang S, Sharp G, Chen G, Mostafavi H, Shirato H (2004) Integrated radiotherapy imaging system (IRIS): Design considerations of tumour tracking with linac gantry-mounted diagnostic x-ray systems with flat panel detectors. Physics in Medicine and Biology 49:243–255
4. Blackall J (2002) Respiratory motion in image-guided interventions of the liver. Ph.D. thesis, Guy's King's and St. Thomas' School of Medicine. King's College, London

5. Brewer J, Betke M, Gierga D, Chen G (2004) Real-time 4d tumor tracking and modeling from internal and external fiducials in fluoroscopy. In: Proceedings of 7th International Conference on Medical Image Computing and Computer Assisted Intervention, LNCS, vol. 3217, pp. 594–601
6. Caselles V, Kimmel R, Sapiro G (1997) Geodesic active contours. International Journal of Computer Vision 1(22):61–79
7. Castellanos M, Lacornerie T, Prevost B, Mirabel X, Lartigau E (2004) Sistemas de contencion de organos en movimiento: Radioterapia adaptada a la respiracion (RAR). In: Alpa Editores (ed.) 3 Jornadas Oncologicas Internacionales. Alpa Editores S.A., Madrid
8. Comaniciu D, Zhou XS, Krishnan S (2004) Robust realtime myocardial border tracking for echocardiography: An information fusion approach. IEEE Transactions on Medical Imaging 23(7):849–860
9. Dieterich S, Tang J, Rodgers J, Cleary K (2003) Skin respiratory motion tracking for stereotactic radiosurgery using the CyberKnife. International Congress Series 1256:130–136
10. Drumond T, Cipolla R (2002) Real-time tracking of complex structures with on-line camera calibration. Image and Vision Computing 20:427–433
11. Elgort D, Duerk J (2004) A review of technical advances in interventional magnetic resonance imaging. Academic Radiology 12:1089–1099
12. Gentile C, Camps O, Sznaier M (2004) Segmentation for robust tracking in the presence of severe occlusion. IEEE Transactions on Image Processing 13(2):166–178
13. Kang D, Kim C, Seo Y, Kweon I (1999) A fast and stable method for detecting and tracking medical organs in MRI sequences. IEICE Transactions on Informatics and Systems E83-D(2):497–499
14. Kass M, Witkin A, Terzopopulos D (1987) Snakes: Active contour models. In: First International Conference On Computer Vision, vol. 1, pp. 259–268
15. Keall P, Mageras G (2004) Managing respiratory motion in radiation oncology. In: AAPM 46th Annual Meeting
16. Keall P, Joshi S, Vedam S, Siebers J, Kini V, Mohan R (2005) Four-dimensional radiotherapy planning for DMLC-based respiratory motion tracking. Medical Physics 32(4):945–951
17. Kocak D, da Vittoria Lobo N, Widder EA (1999) Computer vision techniques for quantifying tracking and identifying bioluminescent plankton. IEEE Journal of Oceanic Engineering 24(1):81–95
18. Lotjonen J (2001) Segmentation of MR images using deformable models: Applications to cardiac images. International Journal of Bioelectromagnetism 3
19. Mageras G, Yorke E (2004) Deep inspiration breath hold and respiratory gating strategies for reducing organ motion in radiation treatment. Seminars in Radiation Oncology 14(1):65–75
20. Malciu M, Preteux F (2000) A robust model-based approach for 3d head tracking in video sequences. In: Proceedings of 4th International Conference on Automatic Face and Gesture Recognition 1, pp. 69–173
21. Malladi R, Sethian JA, Vemuri BC (1995) Shape modeling with front propagation: A level set approach. IEEE Transactions on Pattern Analysis and Machine Intelligence 17(2):158–175
22. McInerney T, Terzopoulos D (1995) A dynamic finite element surface model for segmentation and tracking in multidimensional medical images with application to cardiac 4d image analysis. Journal of Computerized Medical Imaging and Graphics 19:69–83
23. Minohara S, Kanai T, Endo M, Noda K, Kanazawa M (2000) Respiratory gated irradiation system for heavy-ion radiotherapy. International Journal of Radiation Oncology 47(4):1097–1103
24. Mukherjee D, Ray N, Acton ST (2004) Level set analysis for leukocyte detection and tracking. IEEE Transactions on Image Processing 13(4):562–572
25. Murphy M (2004) Tracking moving organs in real time. Seminars in Radiation Oncology 14(1):91–100
26. Nederveen A (2002) Image guided position verification for intensity modulated radiotherapy of prostate cancer. Ph.D. thesis, University Utrecht

27. Osher S, Fedkiw R (2001) Level set methods: An overview and some recent results. Journal of Computation Physics 169(2):463–502
28. Paragios N (2002) A level set approach for shape-driven segmentation and tracking of the left ventricle. IEEE Transactions on Medical Imaging 22(6):773–776
29. Paragios N, Deriche R (1997) A PDE-based level-set approach for detecting and tracking of moving objects. Rapport de Recherche. INRIA Sophia Antipolis 1(3173):1–29
30. Paragios N, Deriche R (1999) Geodesic active regions for motion estimation and tracking. Tech. rep., Institut National de Recherche en Informatique et en Automatique
31. Pardas M, Sayrol E (2001) Motion estimation based tracking of active contours. Pattern Recognition Letters 22:1447–1456
32. Rife J, Rock SM (2003) Segmentation methods for visual tracking of deep-ocean jellyfish using a conventional camera. IEEE Journal of Oceanic Engineering 28(4):595–608
33. Sethian J (1999) Level Set Methods and Fast Marching Methods. Cambridge University Press, Cambridge
34. Sharp G, Jiang S, Shimizu S, Shirato H (2004) Prediction of respiratory tumour motion for real-time image-guided radiotherapy. Physics in Medicine and Biology 49:425–440
35. Shirato H, Seppenwoolde Y, Kitamura K, Onimura R (2004) Intrafractional tumor motion: Lung and liver. Seminars in Radiation Oncology 14(1):10–18
36. Tang J, Dieterich S, Cleary K (2004) Respiratory motion tracking of skin and liver in swine for CyberKnife motion compensation. In: Proceedings of SPIE Medical Imaging, vol. 5367, pp. 729–734
37. Tsechpenakis G, Rapantzikos K, Tsapatsoulis N, Kollias S (2004) A snake model for object tracking in natural sequences. Signal processing: Image Communication 19(3):219–238
38. Udupa J, LeBlanc V, Zhuge Y, Schmidt H, Imielinska C, Hirsch B, Woodburn J (2004) A framework for evaluating image segmentation algorithms. Tech. rep., University of Pennsylvania
39. Vauhkonen M, Karjalainen P, Kaipio J (1998) A Kalman filter approach applied to the tracking of fast movements of organs boundaries. In: Proceedings of the 20th Annual International Conference of the IEEE Engineering in Medicine and Biology Society 2, pp. 1048–1051
40. Wong K, VanMeter J, Fricke S, Maurer C, Cleary K (2004) MRI for modeling of liver and skin respiratory motion. Computer Aided Radiology and Surgery
41. Yahia-Chereif L, Gilles B, Molet T, Magnenat Thalmann N (2004) Motion capture and visualization of the hip joint with dynamic MRI and optical systems. Computer Animation and Virtuals Worlds 15:377–385
42. Zeng R, Fessler J, Balter J (2005) Respiratory motion estimation from slowly rotating x-ray projection: Theory and simulation. Medical Physics 32(4):984–991

Human Hand Kinematic Modeling Based on Robotic Concepts for Digit Animation with Dynamic Constraints

Bertrand Tondu[1]

Abstract The recent development of highly anthropomorphic avatars in computer graphics has emphasized the importance of accurate hand kinematic models. Although kinematic methods derived from robotics have recently been applied to the modeling of hands, we consider that original/new and relevant results can be brought into play with the use of more advanced applications of robotic techniques to human hand kinematic modeling. Our chapter analyses some of these questions both in the non-differential and differential fields. More specifically, we study how to integrate the peculiar natural digit movement constraints into robotics-based inverse kinematic modeling. As a result, we propose an original approach based on an interpretation of each joint dynamic constraint as a linear joint synergy. This leads to defining the considered digit as a serial chain kinematically redundant in position and reducing the dimension of its joint space by associated joint synergies. The method is applied to the Cartesian positioning simulation of a 4 d.o.f. index model; a comparison with a Jacobian pseudo-inverse-based approach emphasizes its relevance.

Introduction

Progress recently made in three-dimensional computer graphics for animating complex avatars has led to a renewed interest in the design of accurate human hand kinematic models. The role of these advanced articulated hand models is twofold: first, hand animation greatly contributes to the global-human-gesture-mimicking of the avatar; second, applications of gesture recognition and social communication limit their graphical model to a hand moving in space. Accurately articulated hand models are thus essential for a non-ambiguous communication between the virtual hand and its real interlocutor.

---

[1] LATTIS, University of Toulouse, Campus de Rangeuil, 31077 Toulouse, France
Bertrand.Tondu@insa-toulouse.fr

N. Magnenat-Thalmann et al. (eds.), *Recent Advances in the 3D Physiological Human*,
DOI 10.1007/978-1-84882-565-9_4,

Kinematic hand modeling is a delicate problem combining as it does the difficulties of modeling a vast array of physiological joints with the need for a rigorous mathematical tool able to integrate the so-called static and dynamic constraints peculiar to human hand motions. To cope with this problem the present chapter proposes an analysis of the application of robotic tools to human hand kinematic modeling. Because we limit our study to the purely kinematic issue on purpose, we will not consider the dynamic problem stemming from taking into account both the skeletal and muscular systems as recently done [1, 2, 3].[2] This point of view is motivated by the possibility to separate the kinematic and dynamic models inside a hierarchy of motor control as considered in robotics [4]. Robot kinematic modeling tools have already been applied to hand modeling but, in our opinion, without always using all their mathematical possibilities to full advantage. In particular we wish to demonstrate the possibility of expressing linear hand motion dynamic constraints by means of robotic-based Jacobian digit matrices. The next section discusses the kinematic modeling of the human hand as a branched chain, and the third section deals with the application of Denavit–Hartenberg tables to hand kinematic modeling, in association with the so-called 4×4 Denavit–Hartenberg homogeneous matrices. The fourth section develops an original joint synergy approach that is applied to a digit serial kinematic chain including linear dynamic constraints.

Kinematic Modeling of the Human Hand Based on Robotics Concepts

The Hand as a Branched Kinematic Chain

The hand is clearly the most complicated articulated set of the human skeleton. This complexity can, however, be hierarchically placed into two kinematic groups: (i) fingers and thumb and (ii) carpus. The kinematic modeling of the carpus is particularly difficult due to its "geometrically variable" structure viewed by Kapandji as a "bag of marbles rather than a single structure" [5, p. 154]. The eight bones constituting the proximal row (scaphoid, lunate, triquetrum, pisiform) and the distal row (trapezium, trapezoid, capitate, hamate) are joined by complex articulations difficult to assimilate to mechanical-like joints. It is theoretically possible to assimilate each intercarpal joint by a 3 d.o.f. planar-type mechanical joint, as illustrated in Fig. 1a, but how to control the resulting global mobility in a joint-physiology mimetic way? A simple method to overcome this difficulty consists of

---

[2] Note that the 'dynamic' term in the 'hand dynamic constraints' expression has a kinematic meaning as it will appear in the chapter.

modeling the hand as a branched structure whose root is the wrist. According to this approach, the articular complex of the wrist containing the radiocarpal joint between the radial head and the proximal row of the carpus, followed by the mid-carpal joint between the proximal and distal rows of the carpus, is modeled as a single joint whose mobility in abduction–adduction and flexion–extension is globally defined. As a consequence, the carpus can be considered as a rigid structure playing the role of hand-base. Standard stick-schemes of the hand skeleton, initiated by Buchholz and Armstrong [6], as illustrated in Fig. 1b's recent example, are based on this principle.

If besides, we consider that "the second and third metacarpals are linked to the trapezoid and capitate and to each other by tight-fitting joints that are basically immobile" [8, p. 378], it is possible to define the hand base as the hand "immobile unit" associating these metacarpals to the carpus, as shown in Fig. 1c.

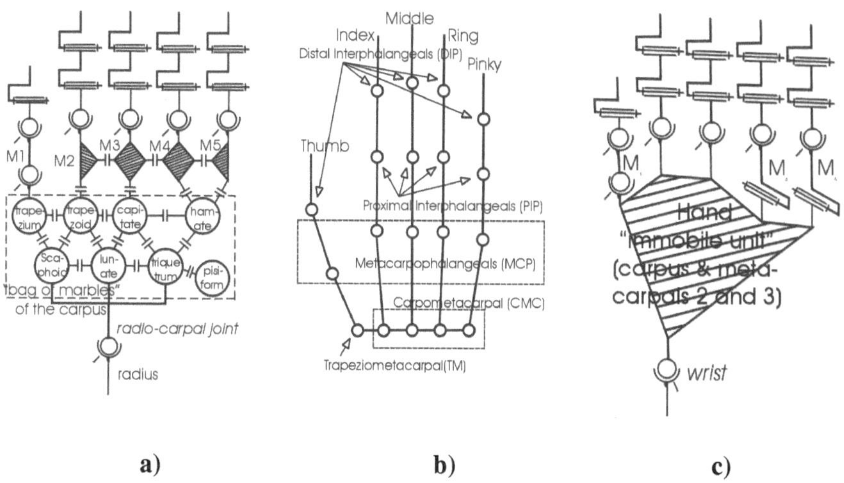

Fig. 1. Attempt of a mechanical-like representation of the hand skeleton. (**a**) Highlighting of the carpus kinematic complexity (see text). (**b**) Usual stick-scheme of the hand skeleton in which each circle corresponds to a physiological joint (redrawn from [7]). (**c**) Simplified representation in which the carpal bones, M2 and M3, are considered as a single rigid "immobile unit".

In the latter model, the CMC joint of the thumb, like all MCP joints, has been modeled by 2 d.o.f. universal joints and CMC joints of M4 and M5, as IP, DIP, PIP joints, by 1 d.o.f. revolute joints, according to a proposed mechanical interpretation discussed elsewhere [9].[3] If the interpretation of the interphalangeal joints

[3] Throughout this chapter we use the following abbreviations: the five metacarpals are noted M1 to M5 where the number refers to the digit number (from 1 for the thumb to 5 for the little finger); the proximal, middle and distal phalanges are, respectively, noted PP, MP and DP; CMC stands for carpometacarpal (we will use it also for the thumb as equivalent to TM for trapezometacarpal), MCP for metacarpophalangeal, IP for interphalangeal, DIP for distal interphalangeal and PIP for proximal interphalangeal.

by mechanical revolute joints is relatively obvious, the mechanical interpretation of CMC joints is open to discussion. First, a 5 d.o.f. mobility is generally admitted for the thumb [10] with a 2 d.o.f. TM joint modeled by a 2 d.o.f. universal joint, but by so doing the associated M1 pronation [11] is disregarded. Second, the modeling of M4 and M5 MCP joints is made difficult by the complexity of the palm-hollowing mechanism [5]. For example, 2 d.o.f. joints are considered for their modeling in the VRS programme SANTOS avatar [12].

Whatever the choices made concerning physiological joint modeling, in computer graphics, as in robotics, the Denavit–Hartenberg notation associated with 4×4 homogeneous matrices appear to be the most powerful tool for developing a mathematical model in correspondence with the chosen kinematic model. The passage from a multiple-d.o.f. joint hand kinematic model to a "robotic-based" hand model, however, implies two points:

- Specifying a zero-reference position in joint space. In order to do so, we propose the consideration, in comparison with hand skeleton stick schemes, of an alternative approach, inspired from industrial robotics, in which each link is represented by polyhedrons as illustrated in Fig. 2. The selected zero-reference position corresponds to the "anatomical position" for the fingers and to the rest position for the thumb. This representation mode has, in our view, a second advantage: the specification of any digit end-frame can be easily represented, as long as we limit this choice to a free face of the considered phalange polyhedrons (see Fig. 3b) and advance its mathematical expression by means of homogeneous matrices.

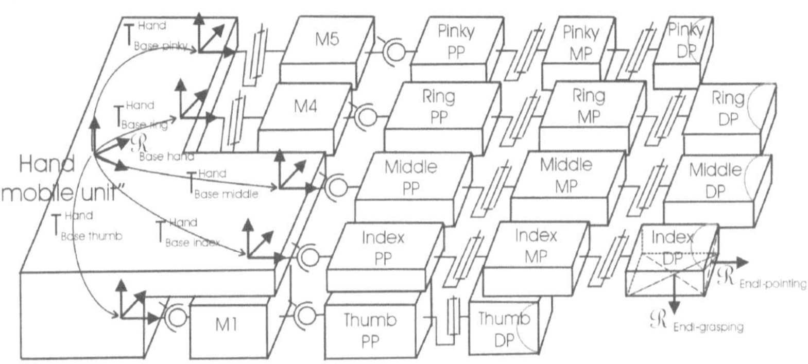

Fig. 2. Alternative kinematic representation of the hand: digit mobile bones are represented like simplified polyhedrons whose free faces can content end-frames (a left hand has been chosen to highlight the thumb zero-position).

- Interpreting each 2 d.o.f. universal joint as a sequence of two successive 1 d.o.f. revolute joints by considering, for example, an abduction–adduction movement followed by a flexion–extension movement, as shown in Fig. 3a. It is important to note that, for obvious reasons of simplicity, present-day hand kinematic modeling generally considers only perpendicular and parallel joint axes, but the theory is naturally adapted to non-perpendicular or non-intersecting axes to be in better agreement with physiological data – see, for example, the discussion concerning the relative positioning of thumb joint axes in [10].

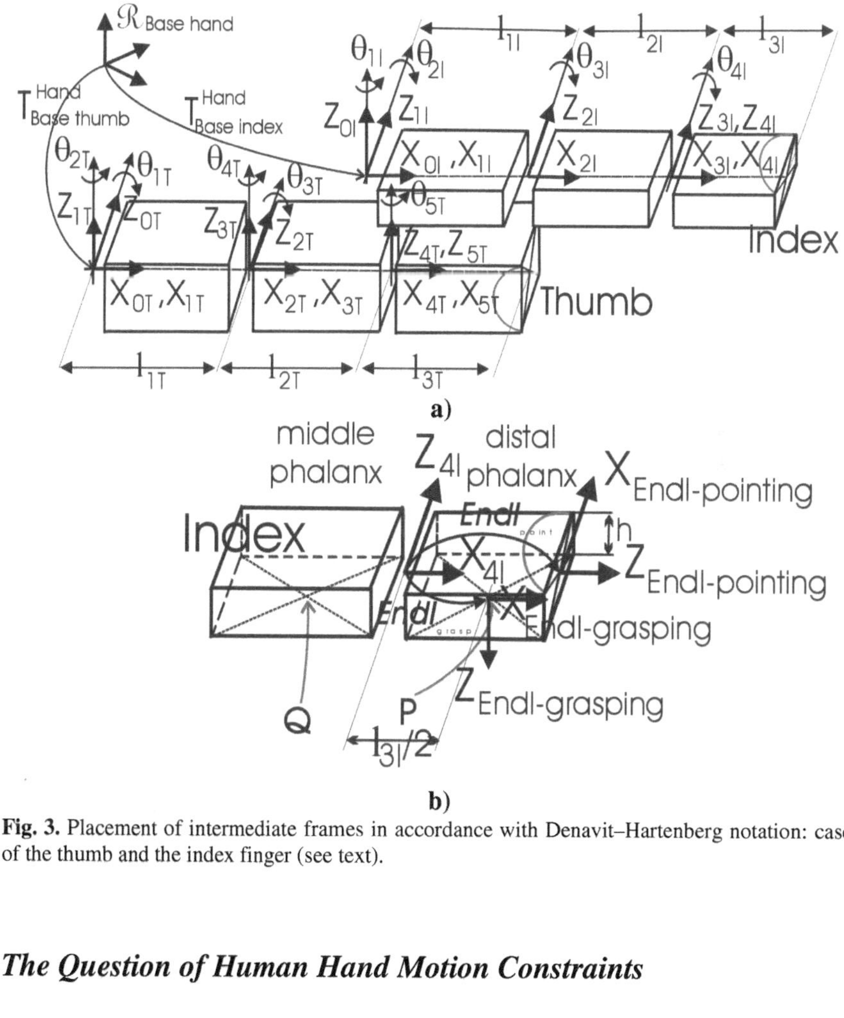

Fig. 3. Placement of intermediate frames in accordance with Denavit–Hartenberg notation: case of the thumb and the index finger (see text).

The Question of Human Hand Motion Constraints

The mobility of the branched kinematic model of the hand is generally determined by considering all joints independently; it is equal, for example, to 23 in the Fig. 1c scheme. But it is well known that in human movements "there is a motor abundance", according to Latash's expression [13]. This motor abundance is used by the central nervous system to define motor synergies adapted to the performance of a type of task or tasks. These synergies naturally exist in the case of the complex movements made by the hand when it manipulates objects or uses tools [14–16]. They are, however, not completely understood notably due to musculature complexity. In a purely kinematic framework, limiting itself to joints free of actuator considerations, these synergies are part of a more general notion of human hand motion constraints. Generally, hand constraints are divided into two main types: "*static constraints* reflecting the range of each parameter, and *dynamic constraints* reflecting joint angle dependencies" [17, p. 56].

Static constraints refer to the range limits of the digits' joint movements. These limits are well known in hand joint physiology – see, for example, Kapandji [5] – by the characterization of the joint range in separate flexion–extension, abduction-adduction or prono-supination movements. Note that this approach is well adapted to robotic-type kinematic modeling by considering the static constraint $\theta_{i\,\min} \le \theta_i \le \theta_{i\,\max}$, where joint variable θ_i corresponds to an elementary motion in flexion–extension, abduction–adduction or prono-supination, and $[\theta_{i\,\min}, \theta_{i\,\max}]$ its evaluated range. A bound case of this static-type constraint can consist of considering the estimated range $[\theta_{i\,\min}, \theta_{i\,\max}]$, small in comparison with other joints' ranges and as a consequence to neglect the corresponding joint: for example, in some models, the thumb trapezo-metacarpal (TM) joint range in abduction–adduction is considered as very limited, and thus the d.o.f. disappears. Let it finally be mentioned that static constraints can if necessary be considered as interdependent, as was done by Lee and Kunii. According to them, "The joint angle limits of the MP joints depend on those of the neighboring fingers" [18, p. 79].

Dynamic constraints in the fingers of the hand are essentially coordination constraints between joint motions as, for example, the typical linear relationship between DIP and PIP joints for digits 2 to 5 [19]:

$$\theta_{DIP} = (2/3)\,\theta_{PIP} \tag{1}$$

Another typical example concerns the following thumb relationship where "flex" designates a flexion–extension motion and "abd" an abduction–adduction motion between a CMC – or TM – joint and a MCP joint [19] (to be compared with our further proposed Eq. 17):

$$\theta_{MCP}^{flex} = 2(\theta_{CMC}^{flex} - \pi/6) \quad \text{and} \quad \theta_{MCP}^{abd} = (7/5)\,\theta_{CMC}^{abd} \tag{2}$$

It is, however, important to note that even in the limited framework of the joint field, these synergies are still poorly understood, and the proposed above-mentioned simple linear relationships can be criticized: for example, relationship (1) appears to be too rough an approximation of intricate control of the hand in the performance of complex tasks such as the playing of musical instruments [20]. For this reason, Lin et al. [21] recently suggested the introduction of a new distinction between this first class of closed form – and generally considered as linear – dynamic constraints and more subtle synergies giving to human hand movements their naturalness and efficiency in tool use. We will not attempt to take into account this "naturalness constraints" class, it being thought as still being too poorly formalized. We thus limit our study to the consideration of both joint static constraints, usual in robotics, and closed-form joint dynamic constraints much more original in robotics which generally avoid joint coordination for obvious technical-complexity and cost reasons. Before developing our robotic-based joint synergy approach, we first analyse the possibilities given by direct and inverse kinematics for hand modeling.

Direct and Inverse Kinematic Modeling of Robot-Like Human Digits

Because the hand is modeled as a tree with five branches, we can define a general base frame associated to the hand immobile unit and a set of five base frames associated to each digit, as illustrated in Fig. 2. In the spirit of industrial robotics, it is subsequently possible to associate a Denavit–Hartenberg representation to the hand kinematic scheme, consisting of combining the five independent Denavit–Hartenberg tables of each digit. Limiting our presentation to the thumb and the index, the Denavit–Hartenberg parameters corresponding to Fig. 3a's intermediate frames' placement are given in Table 1.

Table 1. Proposed digit Denavit–Hartenberg table (see text): case of the thumb and the index finger.

Finger	link	a_i	α_i	d_i	θ_i
Thumb	M1 Abd	0	$+\pi/2$	0	θ_{1T}
	M1 Flex	l_{1T}	$-\pi/2$	0	θ_{2T}
	PP Abd	0	$+\pi/2$	0	θ_{3T}
	PP Flex	l_{2T}	0	0	θ_{4T}
	DP	0	0	0	θ_{5T}
Index	PP Abd	0	$-\pi/2$	0	θ_{1I}
	PP Flex	l_{1I}	0	0	θ_{2I}
	MP	l_{2I}	0	0	θ_{3I}
	DP	0	0	0	θ_{4I}

It is interesting to note that each Denavit–Hartenberg table digit not only is independent of the initial constant transformation $T^{Hand}_{Base\ thumb}$, $T^{Hand}_{Base\ index}$, … defined in Fig. 3 between the hand "immobile unit" and the digit base frame (R_{0T} and R_{0I} in the case of the thumb and the index, as shown in Fig. 3a) but is also independent of the final constant transformation between the frame attached to the last joint axis and the selected digit end-frame ($R_{EndI\text{-}grasping}$ and $R_{EndI\text{-}pointing}$ in Fig. 3b). This representation mode seems to us preferable to Chua et al. approach [22] which includes the parameters of the initial transformations inside the table, or to the recent Dragulescu et al. approach [23] which proposes the inclusion of wrist joint variables in the D-H digit tables. Our opinion is close to the one adopted in the SANTOS hand kinematic model [12], with the difference that we introduce the mention of the movement's type, abduction–adduction or flexion–extension within the D-H table to distinguish it from a table with a reverse order.

We now turn to the possibilities of direct and inverse derived kinematic models. We limit ourselves to the case of 4 d.o.f. fingers such as the index finger. In the following, to simplify our notation, we now omit the digit letter in lengths and joint angles. Although the Denavit–Hartenberg notation is now widely used in hand modeling studies, it is important to note that the associated use of the so-called 4×4 Denavit–Hartenberg homogeneous matrices is not so widespread (as, for example, in [22] or in [24]). We nevertheless think that its use is very efficient, as has been proven in industrial robotics. Using R. Paul's notation [25], we obtain from the D-H sub-table index finger the corresponding D-H matrices A1 to A4, and directly from Fig. 3b the final constant transformation matrices in the two cases considered for pointing, $\boldsymbol{End}_{point}$, or for grasping, $\boldsymbol{End}_{grasp}$:

$$A_1=\begin{bmatrix} C_1 & 0 & -S_1 & 0 \\ S & 0 & C_1 & 0 \\ 0 & -1 & 0 & 0 \\ 0 & 0 & 0 & 1 \end{bmatrix},\ A_2=\begin{bmatrix} C_2 & -S_2 & 0 & l_1C_2 \\ S_2 & C_2 & 0 & l_1S_2 \\ 0 & 0 & 1 & 0 \\ 0 & 0 & 0 & 1 \end{bmatrix},\ A_3=\begin{bmatrix} C_3 & -S_3 & 0 & l_2C_3 \\ S_3 & C_3 & 0 & l_2S_3 \\ 0 & 0 & 1 & 0 \\ 0 & 0 & 0 & 1 \end{bmatrix},$$

$$A_4=\begin{bmatrix} C_4 & -S_4 & 0 & 0 \\ S_4 & C_4 & 0 & 0 \\ 0 & 0 & 1 & 0 \\ 0 & 0 & 0 & 1 \end{bmatrix},\ \boldsymbol{END}_{point}=\begin{bmatrix} 0 & 0 & 1 & l_3 \\ 0 & -1 & 0 & 0 \\ 1 & 0 & 0 & 0 \\ 0 & 0 & 0 & 1 \end{bmatrix},\ \boldsymbol{END}_{grasp}=\begin{bmatrix} 1 & 0 & 0 & l_3/2 \\ 0 & 0 & 1 & h \\ 0 & -1 & 0 & 0 \\ 0 & 0 & 0 & 1 \end{bmatrix} \quad (3)$$

In the case of the "grasping" end-frame choice, we finally obtain the following direct kinematic model in the form of matrix $\boldsymbol{T}_{index}$ relating R_{0I} to $R_{EndI\text{-}grasping}$: $\boldsymbol{T}_{index}=A_1A_2A_3A_4\boldsymbol{End}_{grasp}$ and thus

$$\boldsymbol{T}_{index}=\begin{bmatrix} C_1C_{234} & S_1 & -C_1S_{234} & C_1\,[l_1C_2+l_2C_{23}+(l_3/2)C_{234}-hS_{234}] \\ S_1C_{234} & -C_1 & -S_1S_{234} & S_1\,[l_1C_2+l_2C_{23}+(l_3/2)C_{234}-hS_{234}] \\ -S_{234} & 0 & -C_{234} & -l_1S_2-l_2S_{23}-(l_3/2)S_{234}-hC_{234} \\ 0 & 0 & 0 & 1 \end{bmatrix} \quad (4)$$

Since the finger has only 4 d.o.f., it is clear that the operational location of the distal phalanx can be uniquely defined by means of four parameters. It is, however, seldom noted that these four operational parameters can be chosen in different ways: in our case, a first choice can consist of selecting the three independent position parameters P_x, P_y, P_z defined in Eq. (4) final column and one orientation parameter α, as follows:

$$\begin{cases} P_x = C_1\ [l_1C_2+l_2C_{23}+(l_3/2)C_{234}-hS_{234}] \\ P_y = S_1\ [l_1C_2+l_2C_{23}+(l_3/2)C_{234}-hS_{234}] \\ P_z = -l_1S_2-l_2S_{23}-(l_3/2)S_{234}-hC_{234} \\ \alpha = \theta_2+\theta_3+\theta_4 \end{cases} \quad (5)$$

But it is also possible to define the operational vector by the two position parameters P_y, P_z associated to the two orientation parameters θ_1, α. Moreover, we wish to emphasize the possibility of defining operational variables associated to several digit phalanges, which can be required by some grasping modes. In the case of our 4 d.o.f. index finger, it is thus possible to consider an end-point located on the internal face of the middle phalanx, noted ***Q*** in Fig. 3b, whose components can be directly deduced from product $\boldsymbol{A_1A_2A_3End}_{\text{grasp}}$ where $\boldsymbol{End}_{\text{grasp}}$ is identical to the previous one with l_2 instead of l_3. The fourth operational variable defining the position of the distal phalanx can simply be joint variable θ_4, as illustrated in Fig. 4. It is clear that operational vector choice depends on digit task.

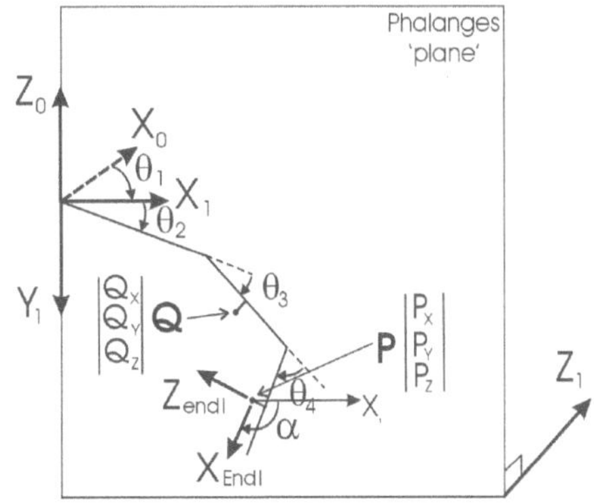

Fig. 4. Multiple possibilities of defining the operational space of a digit – case of the 4 d.o.f. index (see text).

In every case, it is possible to determine a corresponding inverse kinematic model mapping the four-dimensional joint space with the selected four-dimensional operational space. For example, in the case of (P_x, P_y, P_z, α) operational variables we obtain by direct application of the robotic-based inverse kinematic method [25]

$$\theta_1 = \text{atan2}(P_y, P_x)$$

$$\theta_3 = \text{atan2}(\sqrt{1-C_3^2},\ C_3) \quad \text{with} \quad C_3 = (f^2+g^2-l_1^2-l_2^2)/2l_1l_2$$

$$\text{where } f=(C_1P_x+S_1P_y)-(l_3/2)C_\alpha+hS_\alpha \text{ and } g=P_z+(l_3/2)S_\alpha+hC_\alpha$$

$$\theta_2 = \text{atan2}(-l_2S_3f-(l_1+l_2C_3)g\ ,\ (l_1+l_2C_3)f-l_2S_3g)$$

$$\theta_4 = \alpha-\theta_2-\theta_3 \quad (6)$$

The inverse kinematic model can easily take into account static constraints by means of a belonging test of the resulting joint vector generated by Eq. (6) inside corresponding joint limits. Due to the inverse kinematic model based on inverse closed-form trigonometric equations, it does appear, however, to be difficult to integrate dynamic constraints into the method. This is why we now consider a differential approach to the problem.

Joint Synergy Control Applied to Digit Kinematic Models

The digit kinematic models developed above are not redundant since the considered joint space and the operational space both have the same dimension. Our idea is to tackle the problem of digit dynamic constraints as a kinematic redundancy control problem in which the dimension of joint space is greater than the operational space, and dynamic constraints are used to reduce the joint space dimension. Typically the operational space is limited to the end-point digit position with three parameters, and the digit joint space has four or five parameters. We first try to develop a general approach to joint synergy relationship-based control, before applying it to the specific case of digit linear dynamic constraints, and then to suggest a simulation example.

General Framework

We will consider in the general case an n-d.o.f. serial kinematic chain and its associated n-dimensional joint space, as well as an m-dimensional operational space where $m<n$. As a consequence, the considered kinematic structure has a "degree of redundancy" equal to $r=n-m$. We will note $\boldsymbol{q}$ as the joint vector and q_i, $1 \le i \le n$ its components, and $\boldsymbol{x}$ the operational vector. We propose the defining of a ***synergy relationship*** – noted *Syn* – as a dependency relationship between a set of p joint variables $(q_{i_1}, \ldots, q_{i_p})$ as follows:

$$Syn(q_{i_1}, \ldots, q_{i_p}) = 0 \tag{7}$$

In the case of a robot with r degrees of redundancy, we consider r joint synergy relationships which, consequently, reduce the dimension of the joint space from n to m. We obtain from Eq. (7) the following differential relationship:

$$\frac{\partial Syn(\boldsymbol{q})}{\partial q_{i_1}} \dot{q}_{i_1} + \ldots + \frac{\partial Syn(\boldsymbol{q})}{\partial q_{i_p}} \dot{q}_{i_p} = 0 \tag{8}$$

which can be written as follows with $S_i(\boldsymbol{q}) = \partial Syn(\boldsymbol{q}) / \partial q_i$:

$$\sum_{i\in\{i_1,\dots,i_p\}} S_i(\boldsymbol{q})\dot{q}_i = 0 \tag{9}$$

Let us now choose among joint variables $(q_{i_1},\dots,q_{i_p})$ a privileged variable q_k which we suggest is called a ***synergy variable associated*** to the given joint synergy relationship. We obtain from Eq. (9)

$$\dot{q}_k = - \sum_{i\in\{i_1,\dots,i_p\},\, i\neq k} \frac{S_i(\boldsymbol{q})}{S_k(\boldsymbol{q})}\dot{q}_i \tag{10}$$

Note that $c_i(\boldsymbol{q}) = -S_i(\boldsymbol{q})/S_k(\boldsymbol{q})$ for $1\le i\le n$ and $i\in\{i_1,\dots,i_p\}, i\neq k$ and $c_i(\boldsymbol{q}) = 0$ otherwise. Relationship (10) can thus be written as

$$\dot{q}_k = \sum_{1\le i\le n,\, i\neq k} c_i(\boldsymbol{q})\dot{q}_i \tag{11}$$

Now introduce this synergy relationship into the robot differential model. If we note $\boldsymbol{J}_i$ as the ith column of the robot Jacobian matrix, we can define, by elimination of velocity joint variable $\dot{q}_k$, new matrix $\boldsymbol{J}_S$ which can be called ***Jacobian matrix associated to the synergy relationship***:

$$\boldsymbol{J}_S = [(\boldsymbol{J}_i + c_i(\boldsymbol{q})\boldsymbol{J}_k)_{1\le i\le n,\, i\neq k}] \tag{12}$$

This approach can be easily generalized to the taking into account of a set of r synergy relationships by associating an own-synergy joint variable $q_{k1},\dots,q_{kr}$ to each of them in such a way as to make them all different. Then gather them into a vector called ***synergy vector*** and noted $\boldsymbol{q}_S$. The corresponding synergy matrix can now be written as follows:

$$\boldsymbol{J}_S = \left[(\boldsymbol{J}_i + c_i(\boldsymbol{q}) \sum_{k\in\{k_1,\dots,k_r\}} \boldsymbol{J}_k\ )_{1\le i\le n,\, i\notin\{k_1,\dots,k_r\}} \right] \tag{13}$$

Let us define the vector $\boldsymbol{q}_{\bar{S}}$ gathering the joint variables which are not included in the $\boldsymbol{q}_S$ vector. We get

$$\dot{\boldsymbol{x}} = \boldsymbol{J}_S(\boldsymbol{q})\dot{\boldsymbol{q}}_{\bar{S}} \tag{14}$$

Consequently, as long as the $m\times m$ $\boldsymbol{J}_S$ matrix is not singular, we deduce an inverse relationship:

$$\dot{\boldsymbol{q}}_{\bar{S}} = \boldsymbol{J}_S^{-1}(\boldsymbol{q})\,\dot{\boldsymbol{x}} \tag{15}$$

which is to be completed by the set of relationships defining $\dot{\boldsymbol{q}}_S$ as a function of $\dot{\boldsymbol{q}}_{\bar{S}}$. In some ways, this approach controls the redundancy according to a principle in opposition to Seraji's "configuration control approach" which is based on an increasing of the operational space dimension in order to make it equal to the joint space. The proposed synergy approach becomes particularly interesting when coefficients $c_i(\boldsymbol{q})$ are all constants independent of $\boldsymbol{q}$. The column vectors of the synergy Jacobian matrix are, subsequently, linear combinations of the column vectors of $\boldsymbol{J}$. This implies that its determinant can be simply computed from the fundamental property of any determinant as an n-linear alternative form. So in this case we obtain the following relationship in which $K_{i_1 \ldots i_n}$ is a coefficient issued from the products of the c_i between them:

$$\det \boldsymbol{J}_S = \sum_{\substack{(i_1, \ldots, i_n) \in \{1, \ldots, n\} \\ \text{according to the considered synergy}}} K_{i_1 \ldots i_n} \det(\boldsymbol{J}_{i_1}, \ldots, \boldsymbol{J}_{i_n}) \tag{16}$$

These results can be a precious help in determining the "algorithmic singularities" peculiar to the method, as well as for facilitating the obtaining of a closed-form solution to the inverse differential problem.

Application to Hand Digits Dynamic Constraints

The dynamic constraint between middle and distal phalanges, expressed in Eq. (1), is often given as a typical example of simple and obvious hand–finger dynamic constraint. In our approach, this is clearly a linear joint synergy relationship. From a mechanical point of view, its advantage can be interpreted as the possibility it allows of combining two movements with limited joint ranges in the same working plane, with the purpose of increasing its attainable area. Within the framework of our robotic-based joint synergy approach, we propose to generalize this physiological model as follows: let us call "elementary movement" a single abduction–adduction, flexion–extension or prono-supination movement, noted by "elem." Consequently, it is proposed to model a dynamic constraint between p joints of the same elementary movement type by the linear relationship:

$$c_{i_1} \theta_{i_1}^{elem} + c_{i_2} \theta_{i_2}^{elem} + \ldots + c_{i_p} \theta_{i_p}^{elem} = cste \tag{17}$$

where $c_{i_1}, \ldots, c_{i_p}$ are constants.[4] In particular, this approach can be suitable for a finger model including both several abduction–adduction movements as flexion–extension movements, as in the case of the 5 d.o.f. thumb kinematic model. According to a recent study by Li and Tang concerning (the) "coordination of thumb during opposition" [11], the experimentally observed "simultaneous flexion across

[4] A definition of a finger "natural posture" by the linear relationship between joint variables θ_1 and θ_2: $\alpha\theta_1 + \beta\theta_2 + \gamma = 0$ where α, β, γ are constants, can be found in [26]. In comparison with this study, our aims are to propose a general joint synergy linear relationship integrated into a Jacobian-based differential approach.

the CMC – or TM –, MCP and IP joints' (p. 505) could be approximated, roughly speaking it is true, by the following linear joint synergy relationship – where c_1, c_2, c_3, b are constants with possibly $c_3 = 0$:

$$c_1\theta_{CMC}^{flex} + c_2\theta_{MCP}^{flex} + c_3\theta_{IP}^{flex} = b \tag{18}$$

and the fact that "abductors/adductors for the CMC joint are also abductors/adductors for the MCP joint, creating concurrent CMC-MCP joint abduction-adduction" (p. 506) by the other linear joint synergy relationship – where c'_1, c'_2 and b' are constants:

$$c'_1\theta_{CMC}^{abd} + c'_2\theta_{MCP}^{abd} = b' \tag{19}$$

Due to these two relationships, the proposed differential method can be applied with, for example, control variables: θ_{CMC}^{abd}, θ_{CMC}^{flex} and θ_{IP}^{flex}, i.e. $q_S = [\,\theta_{MCP}^{abd}\ \theta_{MCP}^{flex}\,]^T$ and $q_{\bar{S}} = [\,\theta_{CMC}^{abd}\ \theta_{CMC}^{flex}\ \theta_{IP}^{flex}\,]^T$.[5] In the general proposed linear joint synergy Eq. (17), coefficients c_i as the "cste" have to be determined experimentally. It can, however, be noted that in the case of a two-joint relationship the interesting possibility of adapting the set of coefficients to the corresponding joint ranges does exist. Let us consider the two joints θ_k and θ_l of respective ranges $[\theta_{k\,min}, \theta_{k\,max}]$ and $[\theta_{l\,min}, \theta_{l\,max}]$. Synergy relationship

$$\theta_k = \frac{(\theta_{k\,\max} - \theta_{k\,\min})}{(\theta_{l\,\max} - \theta_{l\,\min})} \times (\theta_l - \theta_{l\,\min}) + \theta_{k\,\min} \tag{20}$$

insures the coordinated movement of θ_k and θ_l with a perfect recovery of corresponding joint ranges. Relationship $\theta_{DIP} = (2/3)\,\theta_{PIP}$ can be in relatively good agreement with this two-joint linear synergy relationship. In the case of digits 2 to 5, if no passive extension is considered, we can consider $\theta_{DIP\,\min} = \theta_{PIP\,\min} = 0$; furthermore, maximum flexion values increase from the second to the fifth finger [5] (p. 186): for the PIP joint, from a value of about 100° to a maximum of 135° for the latter, and for the DIP joint from a value slightly less than 90° to a maximum of 90°. If we consider a constant value of 90° for the DIP joint, we obtain, from Eq. (20), relationship $\theta_{DIP} = k\theta_{PIP}$ with k varying from k=(90/100) – case of finger 2 – to k=(90/135)=(2/3) ! – case of finger 5. To illustrate the proposed approach in the case of a 4 d.o.f. index finger with the aim of comparing the result with the one obtained using a standard pseudo-inverse-based approach, we adopt for DIP-PIP a (2/3) joint synergy ratio. In accordance with our section 3 finger

---

[5] This 5 d.o.f. model can if necessary be completed by a CMC pronation movement, and the observed fact that 'flexion and pronation [are] linearly coupled at the CMC joint during opposition' [11] (page 506) can be expressed by a supplementary relationship such as – with three new constants c''_1, c''_2 and b'' : $c''_1\theta_{CMC}^{flex} + c''_2\theta_{CMC}^{pron} = b''$. The three chosen control variables are the same in this 6 d.o.f. thumb model with three linear joint relationships.

model, it can be written as follows: $\theta_4 = (2/3)\theta_3$. Applying our approach, we are led to define vectors $\boldsymbol{q}_S$ and $\boldsymbol{q}_{\bar{S}}$, respectively, by $\boldsymbol{q}_S = \theta_4$ and $\boldsymbol{q}_{\bar{S}} = [\theta_1\ \theta_2\ \theta_3]$. The corresponding synergy Jacobian matrix is so defined from knowledge of the 4R index model Jacobian column vectors, as follows:

$$\boldsymbol{J}_{Sindex} = [\boldsymbol{J}_1\ \ \boldsymbol{J}_2\ \ \boldsymbol{J}_3 + (2/3)\boldsymbol{J}_4] \tag{21}$$

By using Fig. 3b's frame notations, we obtain the following expression of the Jacobian index in frame R_2:

$$ {}^2\boldsymbol{J}_{index} = \begin{bmatrix} 0 & j_{12} & j_{13} & j_{14} \\ 0 & j_{22} & j_{23} & j_{24} \\ j_{31} & 0 & 0 & 0 \end{bmatrix} \text{ with } \begin{cases} j_{12} = j_{13} = -l_2 S_3 - l_3 S_{34} - hC_{34} \\ j_{14} = -l_3 S_{34} - hC_{34} \\ j_{22} = l_1 + l_2 C_3 + l_3 C_{34} - hS_{34} \\ j_{23} = l_2 C_3 + l_3 C_{34} - hS_{34} \\ j_{24} = l_3 C_{34} - hS_{34} \\ j_{31} = l_1 C_2 + l_2 C_{23} + l_3 C_{234} - hS_{234} \end{cases} \tag{22}$$

$$\Rightarrow \quad ({}^0\boldsymbol{J}_{Sindex})^{-1} = \frac{1}{D} \begin{bmatrix} 0 & 0 & j_{12}(j_{23}+(2/3)j_{24}) \\ & & -j_{22}(j_{13}+(2/3)j_{14}) \\ j_{31}(j_{23}+\frac{2}{3}j_{24}) & -j_{31}(j_{13}+\frac{2}{3}j_{14}) & 0 \\ -j_{31}j_{22} & j_{31}j_{12} & 0 \end{bmatrix} {}^2R_0 \quad \text{with}$$

$$D = \det \boldsymbol{J}_{Sindex} = j_{31}\left[j_{12}(j_{23}+(2/3)j_{24}) - j_{22}(j_{13}+(2/3)j_{14})\right] \tag{23}$$

and ${}^2\boldsymbol{R}_0$ directly deduced from matrices $\boldsymbol{A}_1$ and $\boldsymbol{A}_2$ given in Eq. (3). Finally, this gives rise to control equation:

$$[\dot{\theta}_1\ \dot{\theta}_2\ \dot{\theta}_3]^T = ({}^0\boldsymbol{J}_{Sindex})^{-1}[\dot{P}_X\ \dot{P}_Y\ \dot{P}_Z]^T \quad \text{and} \quad \dot{\theta}_4 = \frac{2}{3}\dot{\theta}_3 \tag{24}$$

We give in Fig. 5a a simulation of the method in comparison with the standard approach based on the use of the pseudo-inverse, with an additional joint limits' avoidance criterion (see, for example, [27]). In order to better visualize the successive phalange positions, finger movement is limited to a planar trajectory – thus $\theta_1 = 0$ has been imposed. The end-point is placed on the final phalanx as shown in Fig 5a. The dimensions of the three finger phalanges were chosen from Buchholz et al. classical study [28] estimating, for each digit, the different segment lengths (SL) between corresponding joint centres as a ratio of hand length (HL); in the case of the index finger, the following formulae are given: $SL_{prox.\ phal.} = 0.245 \times HL$, $SL_{middle\ phal.} = 0.143 \times HL$ and $SL_{distal\ phal.} = 0.097 \times HL$ [24]. From the American *Naval Biodynamics Laboratory* biometric study [29], we obtain HL = 19.2 cm (p. 18, for a mid-size aviator) and thus we consider $l_1 \approx 4.7$ cm, $l_2 \approx 2.75$cm, $l_3 \approx 1.85$cm. Moreover, we consider h = 0.5 cm. The task is defined as a straight-line Cartesian trajectory from a first point $\boldsymbol{P}_{dep}$ corresponding to $\theta_2 = 30°$, $\theta_3 = \theta_4 = 0°$, to a final point $\boldsymbol{P}_{arr}$ defined in the Cartesian plane ($\boldsymbol{O}_0$, $\boldsymbol{X}_0$, $\boldsymbol{Z}_0$)

by $[0, -P_{Zdep}]^T$, performed according to a velocity profile with a constant cruising speed and initial–final phases at continuous acceleration (see Fig. 5a). Figure 5b, c gives, respectively, successive positions of the finger phalanges generated by the joint synergy control approach and by the pseudo-inverse-based approach, and Figure 5d, e, respectively, the corresponding joint variables in position, velocity, and acceleration.

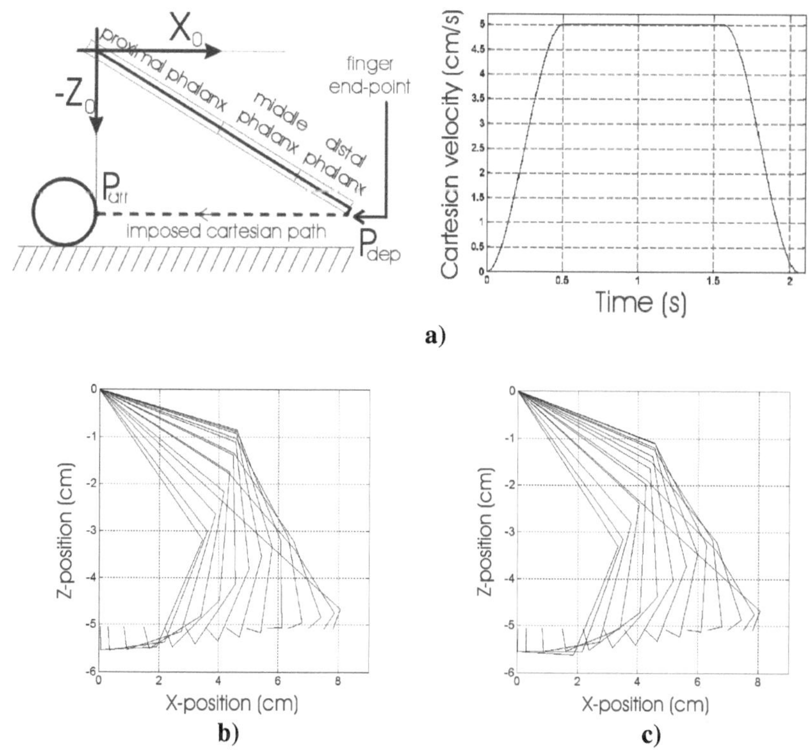

Fig. 5. Comparison between the proposed joint synergy control approach and a classical pseudo-inverse-based method applied to our 4 d.o.f. index finger kinematic model moving on a plane – θ_1 is equal to zero. (**a**) Definition of the task and associated imposed velocity profile. (**b**) Successive positions of the finger phalanges generated by the joint synergy control approach and in c) by the pseudo-inverse-based method.

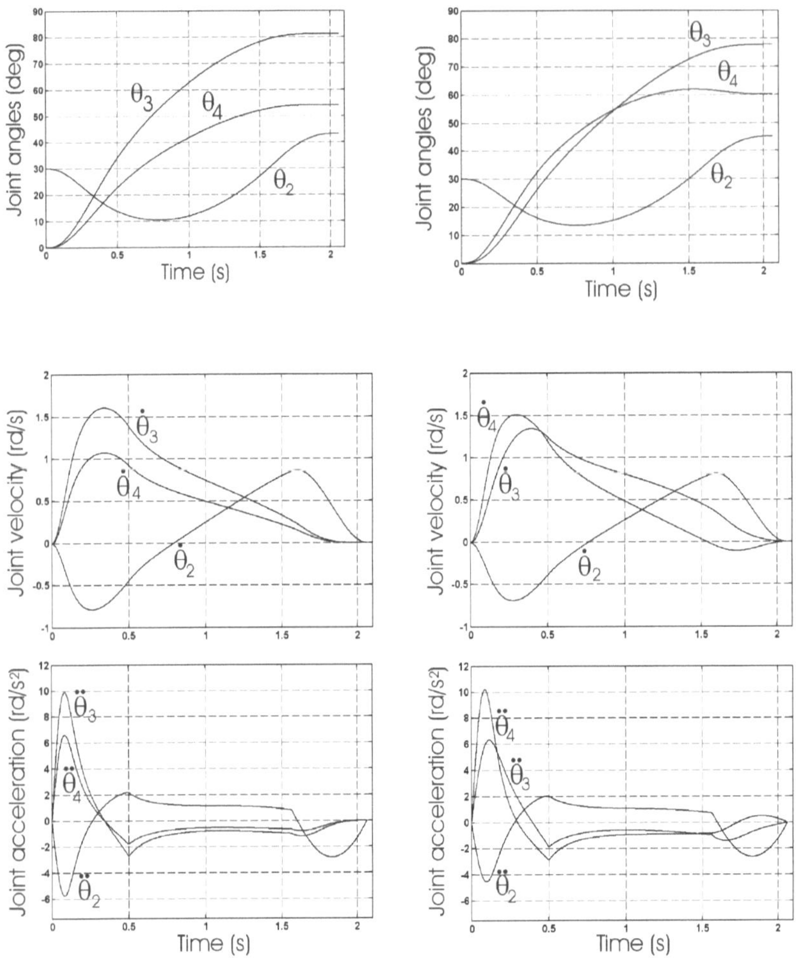

d) **e)**

Fig. 5. (**d**) Corresponding joint positions, velocities, and accelerations in the case of the joint synergy control approach and (**e**) in the case of the pseudo-inverse-based method.

The two results are surprisingly similar concerning maximum values in joint speed and joint acceleration: in comparison with the pseudo-inverse-based method, known to naturally generate the solution at minimum speed norms, the joint synergy control approach imposes a maximum value for $\dot{\theta}_3$ just beyond the greatest joint speed individual value by the pseudo-inverse method.

The proposed approach could be so considered as a relevant alternative to pseudo-inverse-based methods:

- thanks to a judicious choice of synergy linear equations, it leads to a closed-form solution when the pseudo-inverse is generally computed numerically;
- although in principle non-optimal method, it leads to maximum joint velocity and acceleration apparently close to the optimal solution generated by the pseudo-inverse method;
- more fundamentally, it could be a first step towards a new type of robotics-based methods for motion control in operational space integrating partial knowledge of natural movement constraints. The extension of the method to non-linear dynamic constraints is, however, to be studied.

Conclusion

The following points result from our analysis:

1. From the moment an agreement is found to interpret the physiological joints as mechanical joints, the so-called Denavit–Hartenberg parameters are perfectly adapted to a rigorous kinematic modeling of branched chains, such as the hand. However, it is fundamental to remember that the price to pay for this parametrization is a sequencing of elementary flexion–extension, abduction–adduction and pronation movements. Consequently, any corresponding D-H table is limited to a selected sequence which cannot cover all physiological joint possibilities. For example, in the case of the aforementioned CMC–MCP–IP 5 d.o.f. thumb model, four D-H tables are required to take into account all flexion–extension and abduction–adduction combinations.
2. The association of 4×4 D-H homogeneous matrices with initial and constant transformation matrices together elaborates a mapping between the D-H joint space type for each digit and an operational space which can be defined and modified according to the digit task.
3. Unlike static constraints which can easily be tested using a robotic-based inverse kinematic approach, dynamic constraints are difficult to integrate into a closed-form inverse kinematic model. To overcome this disadvantage, we propose a differential approach leading to including linear dynamic constraints into the Jacobian matrix of the digit considered as a redundant serial chain. Preliminary simulations show the relevance of the approach to control the digit in an imposed operational space, even though it satisfies linear joint dynamic constraints.

References

1. Albrecht I, Haber J, Seidel HP (2003) Construction and animation of anatomically based human hand models. In: Proc. of the ACM SIGGRAPH/Eurographics Symposium on Computer Animation, San Diego, California, pp. 98–109
2. Tsang W, Singh K, Fiume E (2005) Helping hand: An anatomically accurate inverse dynamics solution for unconstrained hand motion. In: Proc. of the ACM SIGGRAPH/Eurographics Symposium on Computer Animation, Los Angeles, California, pp. 319–328
3. Sueda S, Kaufman A, Pai DK (2008) Musculotendon simulation for hand animation. In: Proc. of the ACM/SIGGRAPH Int. Conf. on Computer Graphics and Interactive Techniques, Los Angeles, California, article n°83
4. Craig JJ (2008) Introduction to Robotics-Mechanics and Control, 3rd edition, Pearson Prentice Hall, Upper Saddle River, NJ
5. Kapandji IA (2002) The Physiology of Joints', Vol. 1, Upper Limb, 5th edition, Churchill Livingstone, London
6. Buchholz B, Armstrong TJ (1992) A kinematic model of the human hand to evaluate its prehensile capabilities, Journal of Biomechanics 25(2):149–162
7. Wu Y, Huang TS (2001) Hand modeling, analysis, and recognition, IEEE Signal Processing Magazine, May, pp. 51–60
8. Barr AE, Bear-Lehman J (2001) Biomechanics of the Wrist and Hand. In: Basic Biomechanics of the Musculoskeletal System, M Nordin, VH Frankel, Vol. 14, 3rd edition, Lippincott Williams & Wilkins, Baltimore, pp. 358–387
9. Tondu B (2007) Shoulder complex mobility estimation, Applied Bionics and Biomechanics 4(1):19–29
10. Giurintano DJ, Hollister AM, Buford WL, Thompson DE, Myers LM (1995) A virtual five-link model of the thumb, Medical Engineering & Physics 17(4):297–303
11. Li ZM, Tang J (2007) Coordination of thumb joints during opposition, Journal of Biomechanics 40(3):502–510
12. Yang J, Pitarch EP (2004) Kinematic human modeling, Technical Report, SANTOS Virtual Soldier Research Program, The University of Iowa
13. Latash M (2000) There is no motor redundancy in human movement. There is motor abundance, Motor Control 4:259–260
14. Santello M, Flanders M, Soechting JF (1998) Postural hand synergies for tool use, The Journal of Neuroscience 18(23):10105–10115
15. Dounskaia N, Van Gemmert AWA, Stelmach GE (2000) Interjoint coordination during handwriting-like movements, Experimental Brain Research 135:127–140
16. Todorov E, Ghahramani Z (2004) Analysis of the synergies underlying complex hand manipulation. In: Proc. of the 26th Annual Int. Conf. of the IEEE EMBS, San Francisco, CA, pp. 4637–4640
17. Erol A, Bebis G, Nicolescu M, Boyle RD, Twombly X (2007) Vision-based hand pose estimation: A review, Computer Vision and Image Understanding 108:52–73
18. Lee J, Kunii TI (1995) Model-based analysis of hand posture, IEEE Computer Graphics and Applications 15(5):77–86
19. Rijpkema H, Girard M (1991) Computer animation of knowledge-based human grasping, ACM-SIGGRAPH Computer Graphics 25(4):339–348
20. ElKoura G, Singh K (2003) Handrix: Animating the human hand. In: Proc. on the Eurographics/SIGGRAPH Symposium on Computer Animation
21. Lin J, Wu Y, Huang TS (2000) Modeling the constraints of human hand motion. In: Proc. Workshop on Human Motion, Los Alamitos, CA, pp. 121–126
22. Chua CS, Guan H, Ho YK (2002) Model-based 3D hand posture estimation from a single 2D image, Image and Vision Computing 20(3):191–202

23. Dragulescu D, Ungureanu L, Menyhardt K, Stanciu A (2007) 3D active workspace of the human hand shaped end effector. In: Proc. 13th IASTED Int. Conf. On Robotics and Applications, Würzburg, Germany, pp. 76–81
24. Biggs J, Horch K (1999) A three-dimensional kinematic model of the human long finger and the muscles that actuate it, Medical Engineering & Physics 21(9):625–639
25. Paul R (1981) Manipulators: Mathematics, Programming and Control, The MIT Press
26. Yasumuro Y, Chen Q, Chihara K (1997) 3D modeling of human hand with motion constraints. In: Proc. IEEE Int. Conf. on 3-D Digi. Imag. and Mod., pp. 275–282
27. Siciliano B (1990) Kinematic control of redundant robot manipulators: A tutorial, Journal of Intelligent and Robotic Systems 3:201–212
28. Buchholz B, Armstrong TJ, Goldstein SA (1992) Anthropometric data for describing the kinematics of the human hand, Ergonomics 35(3):261–273
29. Naval Biodynamics Laboratory (1988) Anthropometry and Mass Distribution for Human Analogues. Volume I: Military Male Aviators, New Orleans, LA: Naval Biodynamics Laboratory

Part II

Physically Based Simulation

Virtual Pulmonary Valve Replacement Interventions with a Personalised Cardiac Electromechanical Model

Tommaso Mansi[1], Barbara André[2], Michael Lynch[3], Maxime Sermesant[4], Hervé Delingette[5], Younes Boudjemline[6], and Nicholas Ayache[7]

Abstract Pulmonary valve replacement (PVR) is a pivotal treatment for patients who suffer from chronic pulmonary valve regurgitations. Two PVR techniques are becoming prevalent: a minimally invasive approach and an open-heart surgery with direct right ventricle volume reduction. However, there is no common agreement about the postoperative outcomes of these PVR techniques and choosing the right therapy for a specific patient remains a clinical challenge. We explore in this chapter how image processing algorithms, electromechanical models of the heart and real-time surgical simulation platforms can be adapted and combined together to perform patient-specific simulations of these two PVR therapies. We propose a framework where (1) an electromechanical model of the heart is personalised from clinical MR images and used to simulate the effects of PVR upon the cardiac function and (2) volume reduction surgery is simulated in real time by interactively cutting, moving and joining parts of the anatomical model. The framework is tested on a young patient. The results are promising and suggest that such advanced biomedical technologies may help in decision support and surgery planning for PVR.

---

[1] Asclepios, INRIA-Sophia-Méditerranée, Sophia-Antipolis, France
tommaso.mansi@sophia.inria.fr

[2] Asclepios, INRIA-Sophia-Méditerranée, Sophia-Antipolis, France
barbara.andre@sophia.inria.fr
Mauna Kea Technologies, Paris, France
barbara.andre@maunakeatech.com

[3] Siemens AG, CT SE SCR2, Erlangen, Germany

[4] Asclepios, INRIA-Sophia-Méditerranée, Sophia-Antipolis, France
maxime.sermesant@sophia.inria.fr

[5] Asclepios, INRIA-Sophia-Méditerranée, Sophia-Antipolis, France
herve.delingette@sophia.inria.fr

[6] Service de Cardiologie Pédiatrique, Hôpital Necker Enfants Malades, Paris, France
younes.boudjemline@nck.ap-hop-paris.fr

[7] Asclepios, INRIA-Sophia-Méditerranée, Sophia-Antipolis, France
nicholas.ayache@sophia.inria.fr

N. Magnenat-Thalmann et al. (eds.), *Recent Advances in the 3D Physiological Human*,
DOI 10.1007/978-1-84882-565-9_5,

Introduction

Pulmonary valve replacement (PVR) is a pivotal treatment for patients who suffer from chronic pulmonary valve regurgitations. Two PVR techniques are becoming prevalent. On the one hand, percutaneous PVR (PPVR) aims at inserting new pulmonary valves using minimally invasive methods [11]. However, this technique is only possible if the diametre of the right ventricle (RV) outflow tract is lower than 22 mm. On the other hand, a recent surgical approach consists in replacing the pulmonary valves and directly remodeling the RV [16]. The surgeon not only replaces the valves but also intentionally resects the regions of the RV myocardium that are impaired by fibrosis or scars, to reduce RV volume and improve its function. However, there is no common agreement about the postoperative effects of these techniques upon the RV function, which most probably depends on patient pathophysiology. Choosing the appropriate therapy for a given patient remains a clinical challenge.

In this chapter, we explore how a fast computational model of ventricular electromechanics can be combined with image processing algorithms and an interactive surgical simulation platform to simulate the direct postoperative effects of these two PVR therapies. In the last decade, several electromechanical (EM) models of the heart have been proposed to simulate the phenomena that govern the cardiac activity, from electrophysiology to biomechanics [8, 13, 18]. Primarily developed for the understanding of the organ, recent research now aims at applying them in patient-specific simulations [18, 24]. At the same time, platforms for virtual soft-tissue interventions are becoming efficient enough to allow the real-time simulation of complex surgical interventions [4]. In Tang et al. [21], the authors proposed a first promising approach for patient-specific virtual PVR and RV volume reduction surgery. In their study, they made use of an advanced fluid–structure interaction model and considered the myocardium as a passive isotropic tissue. However, although they obtained satisfying results, they did not simulate the preoperative regurgitations and the active properties of the myocardium.

We propose in this study to use an active anisotropic EM model of the heart, with a simple model of regurgitations. As in [18, 21], we aim at controlling the model with clinically related data. Hence, some simplifications in the model are made while still capturing the main features of the cardiac function.

Figure 1 shows the different elements of the proposed framework, made as modular as possible to allow seamless integration of advanced and specific tools. The biventricular myocardium is semi-automatically segmented from clinical 4D cine MRI, excluding the papillary muscles. The mesh obtained from the time frame at mid-diastole is used as 3D anatomical model to simulate patient cardiac function and PVR therapies. The variation of the blood pool volumes throughout the cardiac cycle is computed to calibrate the EM model. After calibration, the EM model is used to simulate the patient cardiac functions. To simulate the replacement of the valves, regurgitations are disabled. Virtual RV volume reduction surgery is interactively performed on the mesh, in real time, using SOFA [3], an open-source soft-tissue intervention platform.

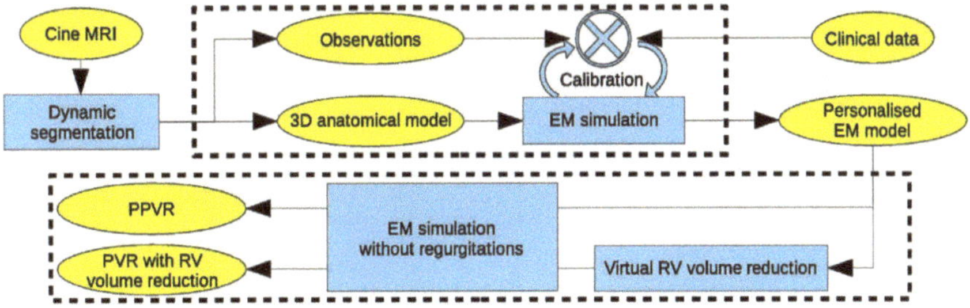

Fig. 1. Pipeline for model-based simulation of PVR therapies (see details in text).

Methods

Anatomical Model

The first step of the framework consists in creating a 3D anatomical model of the patient biventricular myocardium. This model is made up of two main elements, the geometry of the myocardium and the orientation of the myofibres.

Myocardium Geometry

Countless segmentation techniques are available in the literature, often based on prior knowledge [9, 12]. However, because patients with serious pulmonary valve regurgitations present extreme variability in heart anatomy, such methods are unsuitable for our purposes. We thus decided to combine specific image processing techniques to efficiently extract the biventricular myocardium from the cine MRI.

First, the left ventricle (LV) endocardium and the epicardium are delineated on the first frame of the MRI cardiac sequence by using an interactive tool based on variational implicit functions [22]. The user places control points inside, on and outside the desired surface. The algorithm computes in real time the implicit function that interpolates those points and extracts its 0-level set (see Fig. 2). The resulting surface is tracked throughout the cardiac cycle using a diffeomorphic non-linear registration method [23].

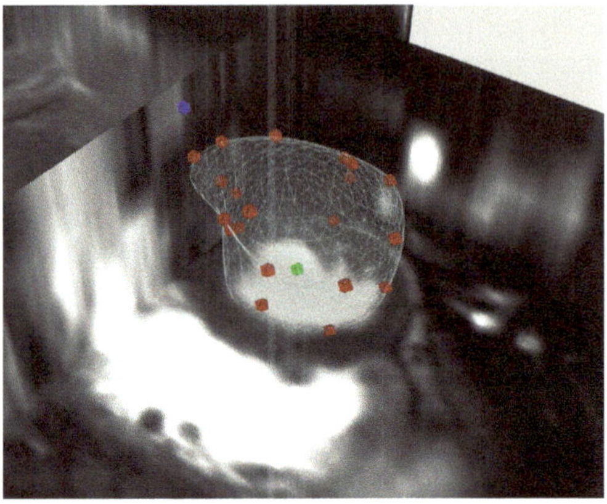

Fig. 2. Interactive segmentation of the left ventricle. The 3D surface (*wireframe*) is interactively modeled by placing surface landmarks (*red points*), inside landmarks (*green point*) and outside landmarks (*blue point*). The algorithm automatically updates the underlying implicit function and extracts the 0-level set surface.

Second, because segmenting the RV in patients with chronic pulmonary valve regurgitations is more challenging than segmenting the LV due to extreme variability in shape and reduced image contrast, the RV endocardium is segmented on all the frames of the cardiac sequence by fitting an anatomically accurate geometrical model. Its position, orientation and scale in the images are determined using minimal user interaction. Boundaries are locally adjusted by training a probabilistic boosting tree classifier with steerable features [25]. The resulting RV endocardium is tracked throughout the cardiac cycle using an optical flow method.

Finally, the binary masks of the segmented epicardium and endocardia are combined together to get a dynamic mask of the biventricular myocardium. The valve plane is manually defined and muscle consistency is ensured by preserving a minimal thickness of 3 mm (mean thickness of a healthy RV myocardium). Dynamic 3D surfaces are computed using CGAL library [1] and simplex-based deformable surfaces [14]. The surface mesh at the mid-diastole time frame is finally transformed into a tetrahedral volume mesh using GHS3D [2] (see Fig. 3).

Myocardium Fibres

Next, the orientations of the myocardium fibres are defined. As their in vivo measurement is still an open challenge, we use a computational model based on observations on anatomical dissections or post-mortem diffusion tensor images. These studies showed that fibre orientation varies from –70° on the epicardium to 0° at mid-wall to +70° on the endocardium [6]. Synthetic fibres are thus created by linearly interpolating their orientation with respect to the short axis plane, from –90° on the epicardium to 0° at mid-wall to +90° on the endocardium. Angles are overestimated to account for averaging, one fibre orientation being associated to one tetrahedron.

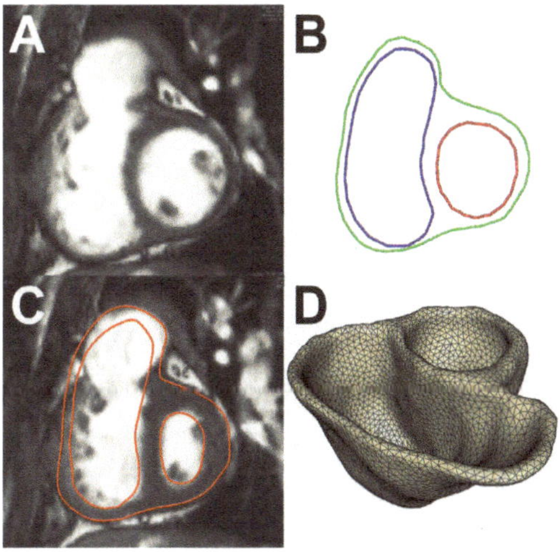

Fig. 3. Cine MRI segmentation. (**a**), (**b**) Image and contours at end-diastole: epicardium (*green*), LV (*red*) and RV (*blue*). (**c**) Tracked myocardium contour at end-systole (*red*). (**d**) 3D mesh of the biventricular myocardium.

Electromechanical Model

Once the volumetric anatomical model is built, computational models of myocardium electrophysiology and biomechanics can be applied on it. Different boundary conditions enable the modeling of the four cardiac phases and the orientation of the myocardium fibres is considered to cope with myocardium anisotropy. For the sake of interactivity, fast algorithms based on finite element method (FEM) are used in this study. Furthermore, because we aim at controlling the model with clinical data, simplifications are introduced while still capturing the main features of the pathological cardiac function.

Electrophysiology

The Eikonal approach [10] detailed in [19] is used to model electrophysiology. This method is computationally efficient and has shown good results when simulating the propagation of the electrical wave in healthy subjects and in patients with no apparent wave re-entry [10, 19], such as in the early stages of the pathologies we are interested in. Essentially, the depolarisation time T_d of the electrical wave is computed at each vertex of the volume mesh by solving the anisotropic Eikonal equation $v^2\left(\nabla T_d^t D \nabla T_d\right) = 1$. In this equation, v is the local conduction velocity and D the tensor defining the conduction anisotropy. In the fibre orientation $\mathbf{f}$ coordinates, D writes $D = \mathrm{diag}(1, \rho, \rho)$, where ρ is the conduction anisotropy ratio between longitudinal and transverse directions.

Biomechanics

Biomechanics are simulated by using the FEM model proposed by Bestel et al. [7], later simplified in [18]. Despite its relative simplicity with respect to other more comprehensive models [8, 13], this model is able to simulate the main features of cardiac motion as observed in images of healthy subjects and patients with less severe diseases [18]. It is controlled by a few clinically related parameters and is fast enough to allow personalisation from clinical data.

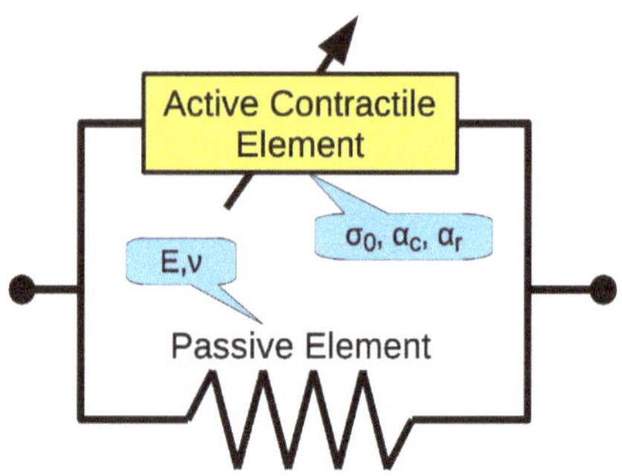

Fig. 4. Scheme of the biomechanical model with related parameters (see details in text).

The constitutive law is composed of two elements (see Fig. 4). A passive element models the properties of the myocardium tissue. We use linear anisotropic visco-elasticity, controlled by the Young's modulus E and the Poisson ratio ν. The second element is an active contractile element, which is controlled by the depolarisation time T_d computed with the Eikonal equation. It generates an anisotropic contraction tensor $\Sigma(t) = \sigma_c(t) \cdot \mathbf{f} \otimes \mathbf{f}$ which results in a contraction force $\mathbf{f}_c$ on each vertex. Σ depends on the time t, the fibre orientation $\mathbf{f}$ and the strength of the contraction $\sigma_c(t)$ as defined by equation (1):

$$\begin{cases} \text{if } T_d \le t \le T_r : \sigma_c(t) = \sigma_0(1 - \exp(\alpha_c(T_d - t))) \\ \text{if } T_r < t < T_d + \text{HP} : \sigma_c(t) = \sigma_c(T_r)(\exp(\alpha_r(T_r - t))) \end{cases} \quad (1)$$

where $T_r = T_d + \text{APD}$ is the repolarisation time, APD the action potential duration, HP the heart period, σ_0 the maximum active contraction and α_c and α_r the contraction and relaxation rates, respectively.

Boundary Conditions and Pulmonary Valve Regurgitations

The four phases of the cardiac cycle – filling, isovolumetric contraction, ejection and isovolumetric relaxation – are simulated as detailed in [18]. During ejection (respectively filling), a pressure constraint equal to the arterial (resp. atrial) pressure is applied to the endocardia. Flows are obtained as the variations of the blood pool volumes. A 3-element Windkessel model [20] is employed to simulate the arterial pressures. During the isovolumetric phases, a penalty constraint is applied to the endocardia to keep the cavity volumes constant. Then, when the ventricular pressure becomes higher (resp. lower) than the arterial (resp. atrial) pressure, ejection (resp. filling) starts.

In this study we are interested in the global effects of the regurgitations on the RV function. Because of the absence of pulmonary valves, blood can always flow between the RV and the pulmonary artery. However, in our implementation of the cardiac phases, regurgitations affect the RV isovolumetric phases only, when the volume can vary due to the regurgitations. If the regurgitation flows ϕ_c and ϕ_r at contraction and relaxation, respectively, are known, by means of echocardiography for instance, we can modify the isovolumetric phases as follows. At each instant t, we first estimate the volume variation ΔV, during Δt, that we would have if the myocardium was free from isovolumetric constraint. Then

- if $|\Delta V| > |\phi_{\{c,r\}} * \Delta t|$, a penalty constraint is applied to each vertex of the RV endocardium such that the resulting volume variation becomes $\Delta V = \phi_{\{c,r\}} * \Delta t$. The effect of the active force $\mathbf{f}_c$ is thus partially compensated.
- Otherwise, no penalty constraint is applied: $\mathbf{f}_c$ is not counterbalanced.

In this way, the blood pool volume can change during the isovolumetric phases according to the measured regurgitation flows.

Real-Time Simulation of Soft-Tissue Interventions

The last component of the proposed pipeline aims at simulating the RV volume reduction surgery. This task is implemented in SOFA, an open-source soft-tissue intervention platform [3]. Interactively, the user remodels the heart geometry by resecting any region of the RV myocardium and closing the remaining wall to recreate the cavity. To cope with large displacements and rotations of the elements, the myocardium biomechanics are simulated using a corotational FEM model [15]. An implicit solver is used to update the mesh position.

Tissue Resection

Tissue resection is performed in two steps. First, a sphere of interest is defined in the SOFA scene by picking two elements of the myocardium surface that define the centre and the radius. Next, all the elements of the mesh lying within this sphere of interest and that are connected to the central element are removed. To this aim, the method proposed by André and Delingette [5], available in the SOFA platform, is used. To ensure interactivity and real-time execution, the authors store all the indices of the mesh elements and all the information attached to them into arrays with contiguous memory storage and short access time. In this way, when the mesh is locally modified, the time to update the data structures does not depend on the total number of mesh elements but only on the number of modified elements. On the other hand, this also implies element renumbering to keep the arrays contiguous in case of element removal, but this procedure is transparent to the user and does not affect the interactivity in our application. Through iterative selection of central elements and radius sizes, the myocardium tissue can be resected as desired (Fig. 5, left panel).

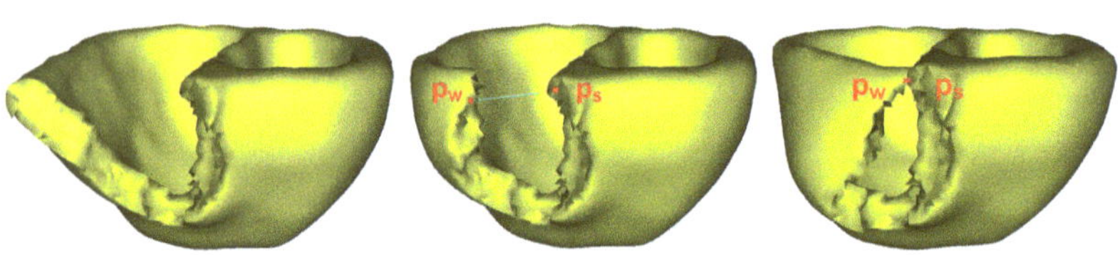

Fig. 5. Virtual suture of the RV. Each side of the resected area (*left panel*) is brought close to each other (*mid panel*). Then, two vertices $\mathbf{p}_w$ and $\mathbf{p}_s$ are picked up and $\mathbf{p}_w$ position is constrained to coincide with $\mathbf{p}_s$, deforming the RV free wall accordingly (*right panel*).

Tissue Attachment

After resection, the RV is interactively and carefully reconstructed without deforming the LV. The boundaries of the resected area are interactively drawn close to each other (Fig. 6). Next, a vertex on one side of the resected area is picked up and fixed to a second vertex on the other side (Fig. 5, mid panel). As a result, the RV morphology globally deforms until reaching a minimal energy state. As this procedure can create holes and generate a chaotic mesh topology in the junction, the deformed geometry is remeshed after rasterisation as a binary image.

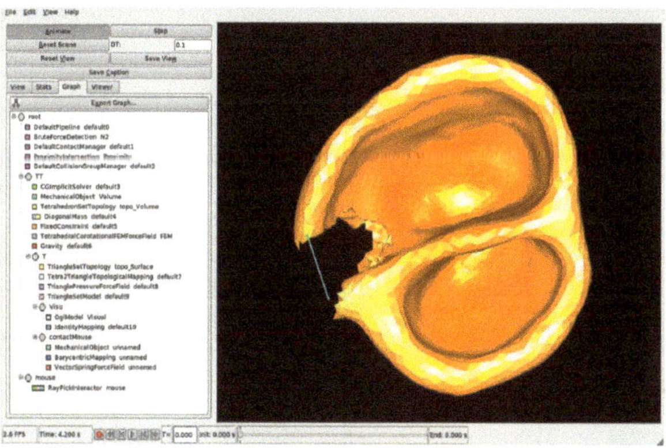

Fig. 6. Screenshot of the SOFA platform. The user is remodeling the geometrical model of a heart. After resection, the user is closing the free wall by pulling it close to the septum (*blue line*).

Myocardium Fibres Recovery

As fibre orientations are intrinsic to the elements of the tetrahedral mesh, they must be preserved during the virtual intervention. Fibres are mapped from the original model to the deformed geometry using local barycentric coordinate systems, preserving in this way their relative orientation in the tetrahedron coordinate system. Then, the resulting fibre orientations are transferred to the final remeshed model by using an intermediate rasterisation of the fibre directions as a vector image.

Experiment and Results

The framework is tested on a randomly selected patient (age 17) with repaired Tetralogy of Fallot, a severe congenital heart disease that requires surgical repair in infancy. The clinical evaluation of this patient showed chronic pulmonary valve regurgitations, an extremely dilated RV and an abnormal motion of the RV outflow tract. LV and RV functions were normal as well as electrophysiology. RV pressure at end-systole was about 50 mmHg (estimated using echocardiography). To date, this patient has undergone no PVR therapy.

Magnetic resonance imaging was performed using 1.5 T MR scanner (Avanto, Siemens Medical Systems, Erlangen, Germany). Retrospective gated steady-state free precession (SSFP) cine MRI of the heart was acquired in the short-axis view covering the entirety of both ventricles (10 slices; slice thickness, 8 mm; temporal resolution, 25 frames). Images were made isotropic (1×1×1 mm) and contrast was enhanced by clamping the tails of the grey-level histogram.

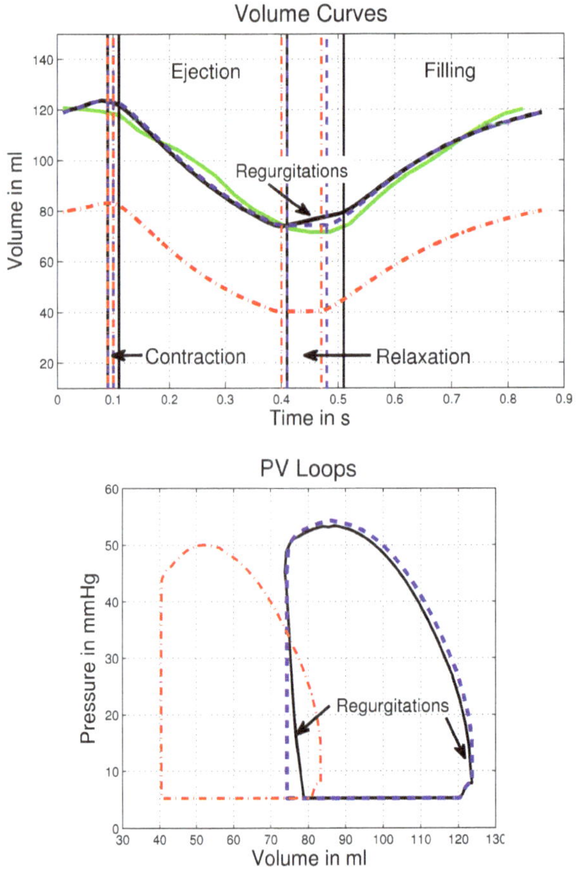

Fig. 7. RV volume curves and pressure–volume loops: (*solid green*) from segmentation; (*solid black*) preoperative simulation; (*dashed blue*) simulation after PPVR; (*dash-dotted red*) simulation after PVR with volume reduction. Vertical bars delineate the simulated cardiac phases. Regurgitations are visible during the isovolumetric phases, when volume should stay constant. As expected, RV volume reduction resulted in a shift of the curves.

Anatomical Model

Visual assessment of the dynamic segmentation showed good agreement (Fig. 3) despite the extremely dilated RV outflow tract. Table 1 provides the ejection fractions (EF) computed from the segmentation. These values will be used as reference when calibrating the EM model. RV volume curve is illustrated in Fig. 7 (*green curve*). The tetrahedral 3D model used for the simulation was made up of 65,181 elements and the dyskinetic region identified on the images was manually delineated (Fig. 8a). To improve the numerical stability of the simulation, the myocardium was artificially thickened by dilating the epicardium.

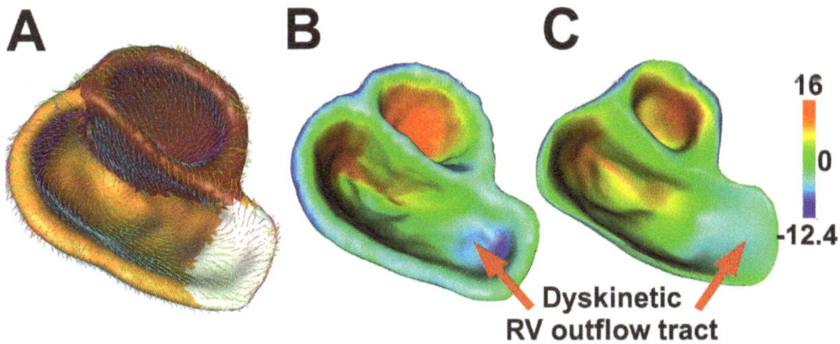

Fig. 8. (**A**) 3D anatomical model at mid-diastole: (*red*) LV, (*orange*) RV, (*white*) observed dyskinetic area, (*colour lines*) fibre orientations. (**B**), (**C**) Radial displacements (in mm) between end-diastole and end-systole (positive values along the outer normal): (**B**) end-systolic shape computed using the model; (**C**) end-systolic shape extracted from the images. Similar colour patterns between (**B**) and (**C**) confirm that the simulated model (**B**) was able to exhibit in this patient realistic motion patterns, in particular the dyskinetic outflow tract.

EM Model Adjustment

Electrophysiology

Since there was no visible anomaly in electrophysiology, electrophysiology parameters were set as in a healthy heart ($v = 500\ \mathrm{ms}^{-1}$, $\rho = 3$). Simulation and observations were time synchronised using the beginning of RV systole. Table 2 reports the computation time of the electrophysiology simulation, without biomechanics.

Biomechanics

Passive biomechanical properties were set according to values reported in the literature: Poisson ratio was set to ensure near-incompressibility of the myocardium ($\nu = 0.48$) and the ratio between fibre stiffness and cross-fibre stiffness was set to 3. Active biomechanical properties were manually personalised to the patient cardiac function. This was possible since a full cycle simulation only took about 15 minutes on a 2.4 GHz Intel Core 2 Duo computer with 4 GB of memory (Table 2). Finally, regurgitation flows ϕ_c and ϕ_r were estimated from the Doppler data ($\phi_c = \phi_r \approx 50\ \text{ml.s}^{-1}$).

Manual personalisation was performed as follows. We first started with reported values for healthy hearts ($\sigma_0 = 100\ \text{kPa.mm}^{-2}$; $\alpha_c = 10\text{s}^{-1}$; $\alpha_r = 20\text{s}^{-1}$). Then, the contractile element of the dyskinetic area was disabled to reproduce the abnormal motion of this region and, through a trial-and-error strategy, we manually adjusted the parameters for both ventricles to simulate the observed cardiac function. The manual adjustment was iteratively performed until the resulting simulated cardiac motion qualitatively coincided with the cine-MR images. The final parameters were: $\sigma_{0_{LV}} = 100\ \text{kPa.mm}^{-2}$; $\sigma_{0_{RV}} = 70\ \text{kPa.mm}^{-2}$; $\sigma_{0_{dysk}} = 0\ \text{kPa.mm}^{-2}$; $\alpha_c = 10\text{s}^{-1}$; $\alpha_r = 10\text{s}^{-1}$. We thus found a normal LV function whereas the RV contractility was weaker, probably because of its dilated morphology and possible fibrosis.

Preoperative Simulation

Realistic EF (Table 1), volume variations (Fig. 7, *black solid curves*) and RV systolic pressures (≈ 54 mmHg) were obtained. Simulated normal displacements of the mesh vertices from their mid-diastole position were locally consistent with those computed from the segmentation (Fig. 8B, C). In particular, the simulated motions of the dyskinetic area, the RV septum and the LV were similar to those estimated from the segmentation. This confirms that the EM model was effectively calibrated and provided, for this patient, realistic motion patterns. Finally, despite our simple regurgitation model, the simulated pressure–volume (PV) loop was consistent with measurements in ToF reported in the literature [17] (these data were not available for this patient).

Table 1. Ejection fractions (EF) computed from the segmentation and simulations.

	LVEF (%)	RVEF (%)
Segmentation	61	41
Simulation: Preoperative	59	40
Simulation: PPVR	59	40
Simulation: PVR with RV reduction	63	51

Percutaneous PVR Simulation

As mentioned in the introduction, PPVR consists in replacing the pulmonary valves using a minimally invasive procedure. The unique change in the cardiac function just after the intervention is thus the stopping of the pulmonary regurgitations. Hence, PPVR therapy was simulated by disabling the regurgitations in our model. After PPVR, the myocardium efficiency was improved: simulated isovolumetric phases were shorter and the end-systolic pressure higher (Fig. 7, *dashed curves*). However, no significant improvement in the pump function was obtained (Table 1). One possible reason could be that our limited regurgitation model accounts for regurgitations that occur during the isovolumetric phases only. Regurgitated volumes during the other phases may have an impact on the preoperative EF.

Simulation of PVR with RV Volume Reduction

As for PPVR, valve replacement was simulated by disabling the regurgitations. Virtual RV volume reduction was performed as illustrated in Fig. 9. First, the observed dyskinetic area (Fig. 8a) was resected in three steps. As reported in Table 2, interactivity was ensured despite the important number of tetrahedra that were removed. Then, the user interactively closes the free wall. The frame rate of the whole process was about 3 fps. Finally, the postoperative scar was simulated by setting the local conduction velocity v near the surgical junction to $0\ ms^{-1}$.

Table 2. Computation time. Values are obtained with a 2.4 GHz Intel Core 2 Duo, 4 GB of memory.

Electromechanical Simulation of the Heart	
1 heart beat, electrophysiology	≈ 1 minute
1 heart beat, electromechanics	≈ 15 minutes
	Virtual surgery
Frame rate	3 fps
Resection, step 1/3 (3938 elements)	0.73s
Resection, step 2/3 (966 elements)	0.08s
Resection, step 3/3 (1404 elements)	0.12s

After the virtual intervention, RV volume effectively decreased (Fig. 7, *dash-dotted curves*) and RV postoperative EF improved significantly (Table 1). Observe that LV EF also increased, which highlights the close relationship between LV and RV function. Finally, RV systolic pressure slightly decreased mainly because of the scar (nonreported simulations without the scar resulted in unchanged RV systolic pressure).

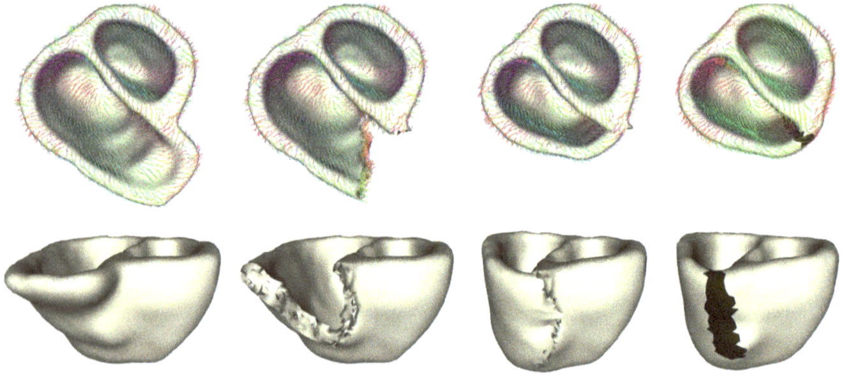

Fig. 9. Virtual RV volume reduction surgery. From *left* to *right*: original mesh, after resection, during attachment, final mesh. Colour lines: fibre orientations. *Black area*: postoperative scar.

Discussion

We have studied the potential of image processing techniques, EM models and virtual soft-tissue intervention platforms to perform virtual and personalised assessment of PVR therapies. The results were promising and suggested that such tools might be used by the clinicians to test different PVR therapies.

A modular framework has been proposed. A simplified but fast EM model was adjusted to the purposes of the study and an interactive soft-tissue intervention platform was adapted. Still, the framework managed to realistically simulate the cardiac function of a randomly selected patient with repaired ToF. For this patient, we found that PVR with RV volume reduction would yield better results than PPVR, just after the intervention. Probable reasons may be the removal of the dyskinetic area and the direct reduction of the RV volume. However, this procedure is very invasive and may be hazardous for the patient, with possible postoperative side effects such as electrophysiological troubles due to the surgical scar. On the other hand, it is worth mentioning that the effects of PPVR are often visible late after replacement. The myocardium adapts itself to its new loading conditions. We thus need to model this phenomenon to simulate the long-term postoperative effects of PPVR.

Based on these observations and owing to the modularity of our framework, more complex models can be used. In particular, future works include improvements of the regurgitation model, to take into account the other compartments, simulation of the long-term effects of the PVR therapies, by considering myocardium remodeling, automated estimation of biomechanical parameters and comprehensive validation using preoperative and postoperative clinical data.

Acknowledgments This work has been partly funded by the European Commission through the IST-2004-027749 Health-e-Child Integrated Project (http://www.health-e-child.org).

References

1. CGAL. http://www.cgal.org
2. GHS3D. http://www-c.inria.fr/gamma/ghs3d/ghs.php
3. SOFA. http://www.sofa- framework.org
4. Allard J, Cotin S, Faure F, Bensoussan PJ, Poyer F, Duriez C, Delingette H, Grisoni L (2007) SOFA – An Open Source Framework for Medical Simulation. In: Medicine Meets Virtual Reality (MMVR'15)
5. André B, Delingette H (2008) Versatile design of changing mesh topologies for surgery simulation. In: International Symposium on Computational Models for Biomedical Simulation – (ISBMS08), pp. 147–156, Springer
6. Arts T, Costa KD, Covell JW, McCulloch AD (2001) Relating myocardial laminar architecture to shear strain and muscle fiber orientation. Am J Physiol Heart Circ Physiol 280(5):2222–2229
7. Bestel J, Clément F, Sorine M (2001) A biomechanical model of muscle contraction. In: Proc. MICCAI 2001, pp. 1159–1161, Springer
8. Hunter PJ, Pullan AJ, Smaill BH (2003) Modeling total heart function. Annu Rev Biomed Eng 5:147–177
9. Kaus MR, von Berg J, Weese J, Niessen W, Pekar V (2004) Automated segmentation of the left ventricle in cardiac MRI. Med Image Anal 8(3):245–254
10. Keener J, Sneyd J (1998) Mathematical Physiology. Springer-Verlag

11. Khambadkone S, Coats L, Taylor A, Boudjemline Y, Derrick G, Tsang V, Cooper J, Muthurangu V, Hegde SR, Razavi RS, Pellerin D, Deanfield J, Bonhoeffer P (2005) Percutaneous pulmonary valve implantation in humans: Results in 59 consecutive patients. Circulation 112(8):1189–1197
12. Lorenzo-Valdes M, Sanchez-Ortiz GI, Elkington AG, Mohiaddin RH, Rueckert D (2004) Segmentation of 4D cardiac MR images using a probabilistic atlas and the EM algorithm. Med Image Anal 8(3):255–265
13. McCulloch A, Bassingthwaighte J, Hunter P, Noble D (1998) Computational biology of the heart: from structure to function. Prog Biophys Mol Biol 69(2–3):153–5
14. Montagnat J, Delingette H (2005) 4D deformable models with temporal constraints: application to 4D cardiac image segmentation. Med Image Anal 9(1):87–100
15. Nesme M, Payan Y, Faure F (2005) Efficient, physically plausible finite elements. In: J Dingliana, F Ganovelli (eds.) Eurographics (short papers), pp. 77–80
16. del Nido PJ (2006) Surgical management of right ventricular dysfunction late after repair of tetralogy of Fallot: Right ventricular remodeling surgery. Semin Thorac Cardiovasc Surg Pediatr Card Surg Annu 9(1):29–34
17. Redington AN, Rigby ML, Shinebourne EA, Oldershaw PJ (1990) Changes in the pressure–volume relation of the right ventricle when its loading conditions are modified. Br Heart J 63(1):45–49
18. Sermesant M, Delingette H, Ayache N (2006) An electromechanical model of the heart for image analysis and simulation. IEEE TMI 25(5):612–625
19. Sermesant M, Konukoglu E, Delingette H, Coudière Y, Chinchapatnam P, Rhode K, Razavi R, Ayache N (2007) An anisotropic multi-front fast marching method for real-time simulation of cardiac electrophysiology. In: Proc. FIMH 2007, pp. 160–169, Springer
20. Stergiopulos N, Westerhof BE, Westerhof N (1999) Total arterial inertance as the fourth element of the windkessel model. Am J Phys Heart Circ Phys 276(1):81–88
21. Tang D, Yang C, Geva, T, del Nido PJ (2007) Patient-specific virtual surgery for right ventricle volume reduction and patch design using MRI-based 3D FSI RV/LV/patch models. In: Proc. CME 2007, pp. 157–162
22. Turk G, O'Brien J (1999) Variational implicit surfaces. Tech. rep., Georgia Institute of Technology
23. Vercauteren T, Pennec X, Perchant A, Ayache N (2007) Non-parametric diffeomorphic image registration with the demons algorithm. In: Proc. MICCAI 2007, pp. 319–326, Springer
24. Wong KCL, Wang L, Zhang H, Liu H, Shi P (2007) Integrating functional and structural images for simultaneous cardiac segmentation and deformation recovery. In: Proc. MICCAI 2007, vol. 4791, pp. 270–277, Springer, Berlin/Heidelberg
25. Zheng Y, Barbu A, Georgescu B, Scheuering M, Comaniciu D (2007) Fast automatic heart chamber segmentation from 3D CT data using marginal space learning and steerable features. In: Proc. ICCV 2007, pp. 1–8

Interactive Simulation of Diaphragm Motion Through Muscle and Rib Kinematics

Pierre-Frédéric Villard[1], **Wesley Bourne**[2], **and Fernando Bello**[3]

Abstract Modeling of diaphragm behaviour is of relevance to a number of clinical procedures such as lung cancer radiotherapy and liver access interventions. The heterogeneity in tissue composition of the diaphragm, as well as the various physiological phenomena influencing its behaviour, requires a complex model in order to accurately capture its motion. In this chapter we present a novel methodology based on a heterogenous model composed of mass–spring and tensegrity elements. The physiological boundary conditions have been carefully taken into account and applied to our model. Thus, it incorporates the influence of the rib kinematics, the muscle natural contraction/relaxation and the motion of the sternum. Initial validation results show that the behaviour of the model closely follows that of a real diaphragm.

Introduction

Diaphragm motion has a crucial influence on surrounding organs. Real-time knowledge of its behaviour is of great significance in the context of various treatments for procedure rehearsal, planning and guidance. We are particularly interested in lung cancer radiotherapy and liver access procedures. In the first case, the lung is subject to the up and down movement of the diaphragm, which can in turn produce tumour movement of up to 5 cm [1]. Liver motion is an important factor to consider in liver access procedures. It has been estimated to be up to 3 cm [2]. Such motion is directly related to diaphragm movement, which is transmitted to the liver through its ligamentous attachments.

[1] Biosurgery and Surgical Technology, Imperial College London, UK
p.villard@imperial.ac.uk

[2] Biosurgery and Surgical Technology, Imperial College London, UK
wesley.bourne@morganstanley.com

[3] Biosurgery and Surgical Technology, Imperial College London, UK
f.bello@imperial.ac.uk

N. Magnenat-Thalmann et al. (eds.), *Recent Advances in the 3D Physiological Human*,
DOI 10.1007/978-1-84882-565-9_6,

A possible solution to predict diaphragm motion is to apply displacement vectors obtained from non-rigid registration in combination with a statistical approach as in Siebenthal et al. [3]. A disadvantage of this approach is that it tends to assume a reproducible breathing cycle, which has been demonstrated to be not always a valid assumption [1]. Our aim is to build a model that can be customized to the patient to comply with the accuracy requirements of radiotherapy. Physically based models of the diaphragm have already been proposed in the past (e.g. [4, 5]), but they are mainly based on mass–spring systems, with their motion dictated by a single parameter.

We proposed in [6] a model that takes into account the heterogeneity of the diaphragm, considering its main tissues types, as well as the relevant physiological phenomena and various forces involved during normal diaphragm movement, thus introducing additional parameters to fine tune/control the motion. It consists of mass–spring elements for the muscular part and tensegrity elements [7] for the tendinous part. Here we focus on the respiration process. Deformations are produced by three different phenomena: the diaphragm contraction, the ribs kinematics and the sternum motion. This chapter presents our methodology and initial validation results on a commercial virtual anatomy model, two radiotherapy patients and a normal volunteer.

Materials and Methods

We start with a careful study of the relevant anatomy and physiology as it is important to understand the real behaviour of the diaphragm in order to be able to establish adequate boundary conditions and select suitable physical models. This section also indicates the integration technique used and the method to monitor the external forces.

Anatomy and Physiology

Research on previous work about diaphragm physiology was focused on its anatomical, histological and mechanical components. The data obtained were analysed and compared to determine viability and validity for its use. Based primarily on the work presented in [8–10], we obtained the main physiological components involved in displacement of the diaphragm and studied how each one of them influences diaphragm motion.

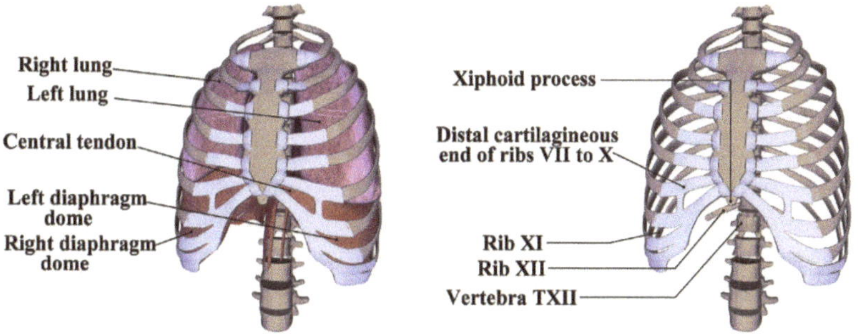

Fig. 1. *Left*: Thorax anatomy; *Right*: Attachment points of the diaphragm.

The thoracic diaphragm is a domed fibromuscular sheet that separates the thoracic and abdominal cavities. It is composed of two domes and the right dome is higher than the left dome (Fig. 1, left). The peripheral part of the diaphragm consists of muscular fibres that converge on a central tendon at the level of the xiphisternum. There are three different groups of muscular fibres: (1) Sternal, attached to the back of the xiphoid process; (2) costal, attached on the inner surfaces of cartilages and adjacent portions of lower six ribs on either side; and (3) lumbar, attached to the lumbar vertebrae (Fig. 1, right).

Given the spatial dimension, topology and tissue composition of the diaphragm, it can be modeled as a heterogenous surface composed of muscle and tendon. The origins of the muscle fibres define the attachment points of the diaphragm. They are used to define the boundary conditions for our model as follows: (1) The xiphoid process of the sternum, (2) the costal margin of thoracic wall (lower six ribs and their costal cartilages), (3) the end of ribs XI and XII, and (4) the vertebrae of lumbar region. We call them the attachment points and the details of the implementation will be described in the following section.

Deformation Modeling

In their survey of deformable models for surgery simulation, Meier [11] claims that mass–spring systems and boundary element models (BEM) are the best models for surgical simulators because they allow a realistic behaviour while supporting near-real-time performance. Nedel and Thalmann [12] proposed a model of muscle based on a surface mass–spring system oriented according to the muscular fibre calling it the "action line". We borrow from this formulation the use of a predefined unique line to parameterize the action of a muscle, as well as the mass–spring formulation, in order to model the muscular component of the diaphragm. Concerning the tendinous part, which can be described as a tough band of fibrous connective tissue, we use a tensegrity model as proposed by Ingber to model certain parts of the body [7].

The word "tensegrity" comes from a contraction of "tensional integrity". It is used to define a mechanical system with components that combine tension and compression in such a way so as to enable the whole system to receive and apply forces, tensions and pressures. Tensegrity systems are composed of two kinds of elements: the elastic elements, which give the system tension, and the rigid elements, which have a constant length and that will exert compression forces.

The evolution of node positions is simulated by taking into account all of its links that could either be elastic or rigid. Each point is represented at a time t by the triplet $(a(t), v(t), p(t))$ representing the acceleration, velocity and position. As we define the forces to apply on the nodes with elastic and rigid links, we can calculate their acceleration with the fundamental law of dynamics:

$$\sum \mathbf{f} = m.\mathbf{a} \tag{1}$$

It is then possible to compute the velocity and the position at time $(t+\Delta t)$:

$$\begin{cases} \mathbf{v}(t+\Delta t) = \int_{t}^{t+\Delta t} \sum \mathbf{f} / m dt + \mathbf{v}(t) \\ \mathrm{p}(t+\Delta t) = \int_{t}^{t+\Delta t} \mathbf{v}(t) dt + \mathbf{p}(t) \end{cases} \tag{2}$$

The elastic links are simulated as in the mass–spring system with two parameters: their initial length $\mathbf{l}_0$ and their elasticity k. A force $\mathbf{F}_e$ exerting on an elastic link to elongate its length from $\mathbf{l}_0$ to $\mathbf{l}$ is thus expressed as follows:

$$\mathbf{F}_e = -\mathrm{k}(\mathbf{l} - \mathbf{l}_0) \tag{3}$$

We add a damping force F_d to systematically and progressively decrease the velocity of each node:

$$\mathbf{F}_d = -\gamma.\mathbf{v} \tag{4}$$

The rigid links in the tensegrity part exert reaction forces to avoid changes in their length. We assume that there is no collision between rigid links. For each time step:

1. The elastic forces $\mathbf{F}_e$ are computed using Eq. (3) as well as the new positions of the nodes as if there were no rigid links.

2. The rigid constrains are applied to ensure that the distance between two nodes remains constant.
3. Given two nodes **A** and **B** linked by a rigid link, **A'** and **B'** are their respective positions after applying the classical mass–spring algorithm (as if there was no rigid link). The real positions $\mathbf{A}_1$ and $\mathbf{B}_1$ considering the rigid links are given by the following:

$$\mathbf{AA1/BB}_1 = \mathbf{B}_1\mathbf{B'/A}_1\mathbf{A'} \quad (5)$$

such that $\mathbf{A}_1$ and $\mathbf{B}_1$ remain on the line **A'B'**.

4. The new velocity is computed at each node.

The resulting set of differential equations is solved using a second-order Runge–Kutta method:

$$\begin{cases} \mathbf{a}(t+\Delta t) = \sum \mathbf{f} \\ \mathbf{v}(t+\Delta t/2) = 1/2\mathbf{a}(t-\Delta t)\Delta t + \mathbf{v}(t) \\ \mathbf{v}(t+\Delta t) = 1/2\mathbf{a}(t-\Delta t)\Delta t + \mathbf{v}(t+\Delta t/2) \\ \mathbf{p}(t+\Delta t) = \mathbf{v}(t+\Delta t) + \mathbf{p}(t) + 1/2\mathbf{a}(t-\Delta t)\Delta t^2 \end{cases} \quad (6)$$

We will now focus on the movement of the diaphragm during inhalation. First, the anterior–posterior diameter increases because the sternum moves forwards as ribs are raised. Second, the transverse diameter increases due to a "pump handle" movement which elevates the ribs. Both movements can be simulated by the combined motion of the ribs, the sternum and the cartilage. We first assume that the ribs are rigid and their motion can be modeled by a kinematics law based on the finite helical axis method as in [13]. The cartilage–sternum follows the motion of the ribs.

The attachment points are modeled by a mass–spring system with a very low elasticity in order to ensure that it remains attached. A fall-off function insures that the deformation field is continuous and smooth. The ones belonging to the diaphragm are automatically obtained such that the distance to the group {ribs, sternum, cartilage} is under a given threshold. Then, the vertical dimension increases due to diaphragm movement during inhalation. This motion can be simulated by the muscle action generation. We define an action line as a vertical line (inferior–superior) passing through given points and with a certain radius of action. The forces of contraction during inhalation or relaxation during exhalation are computed such that their intensity varies according to the distance from the action line.

Results

We now present the purposely written simulation software. We have applied our methodology to a virtual anatomy model and three patients. The focus for the former was on the realism of the motion and especially on the influence of the parameters described previously, whereas for the later we compared our simulated displacements with the real motion measured on a 4D CT scan.

Diaphragm Deformation Framework

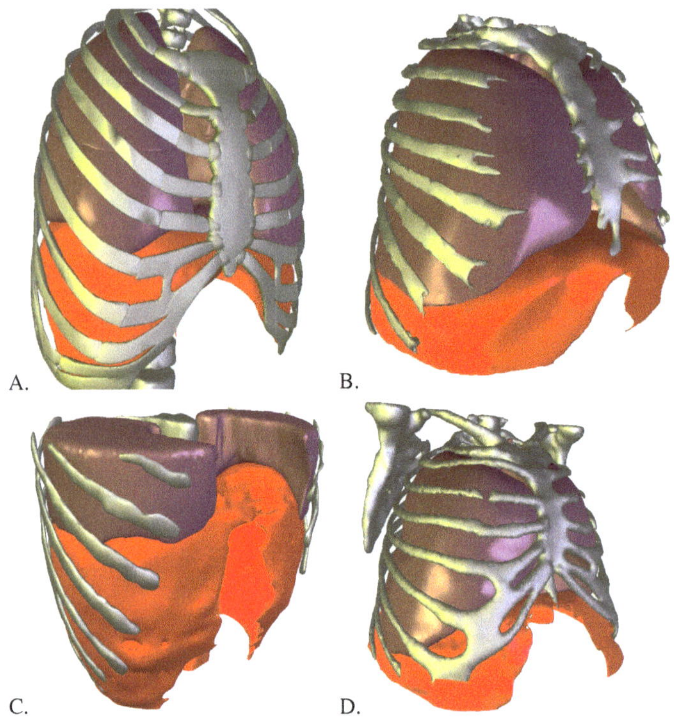

Fig. 2. Interface display of ribcage, lungs and diaphragm: (**A**) model; (B)–(D): patients.

Our framework is coded using Java 3D API.[4] It allows the real-time fine-tuning of simulation parameters in order to study various types of breathing, breath hold, ribs rotation, diaphragm muscle force intensity, etc. The deformation framework enables the visualization of the following elements: (1) external skin by transparency

[4] http://java.sun.com/products/java-media/3D/

(rigid model), (2) the spine (rigid model), (3) the ribs (modeled following a kinematics law), (4) the sternum and the ligaments attached to the ribs (modeled by a very stiff mass–spring system), and (5) the diaphragm (modeled as described above). Figure 2 shows an illustration of the organ deformation display with examples of anatomy: a commercial virtual anatomy model and three real patients. The parameters of the simulation are: damping=–0.02, stiffness= 15, mass point=0.1 and time step=0.01. Results of the simulation and validation using both models are now presented.

Simplified Model

We used a simplified model from a detailed commercial anatomy set (Anatomium 3D by CF Lietzau 3D Special Service). The anatomical parts are shown in Fig. 1. The diaphragm was remeshed to obtain a surface mesh composed of 11,359 nodes and 22,722 triangles. The tendinous tissue and the muscle tissue were separated by a plane. The rigid links of the tensegrity part were created by linking the upper surface to the lower surface with a heuristic algorithm. The reasoning behind it is as follows. Once a suitable distance has been chosen, a list of close masses is found. These masses could either be neighbouring masses on the same surface or ideally those on opposing surfaces. Testing the angle removes those on the same surface, leaving a sorted list of masses on the opposing surface.

Figure 3 is an illustration of the simulation results. The diaphragm geometry at the beginning of the simulation (end of inhale) is represented by a wireframe. The final geometry (end of exhale) is represented by the coloured surface. It can be noticed that the tendon has kept its smooth geometry due to the tensegrity elements, while the muscle part has extended more due to their elastic elements. The influence of the three phenomena mentioned before can be seen here: the muscle relaxation that increases the height of the domes, the ribs rotation that presses the diaphragm on each side and the sternum going backwards that pushes the diaphragm.

Now that the realism of the simulations has been evaluated, we focus on the computational cost of the simulation parameters. The tests below were performed on a dual core 2.4 GHZ machine with 2 GB of RAM and a 256 MB graphics card. The operating system was Windows Vista Service Pack 1 with Java 1.6 installed.

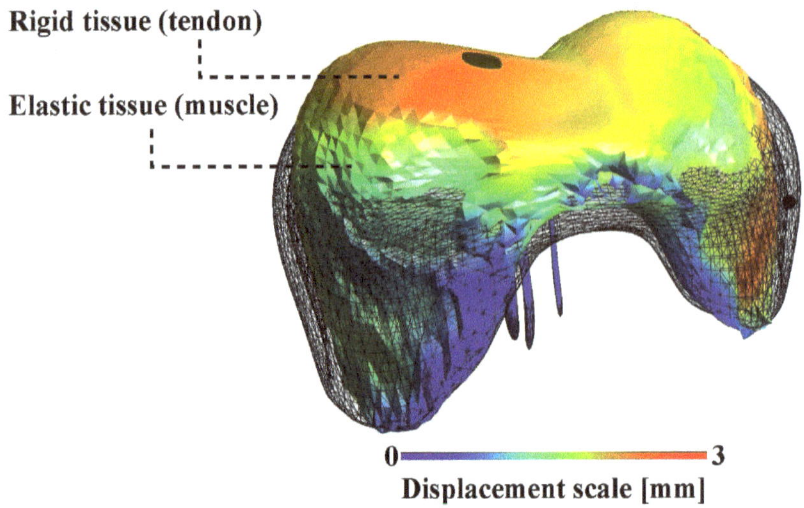

Fig. 3. Illustration of diaphragm motion during inhale.

Table 1 provides a succinct summary of all the data. It gives the average iterations per second after five trials. The relationship between the parameters of the simulation and the mesh size is available. The computational time of the rib kinematics is shown to be quite high as when they are removed in the "Diaph.muscle" trial the average iterations per second hits its peak. Regarding the cost on the speed of the simulator by the addition of tensegrity, it shows that the largest change is visible in the high-quality perfect model. This is due to the very high number of connections created in this case.

Table 1. Iterations per second of all the simulations on the virtual anatomy model. The trials have been averaged into a single value.

Resolution	Vertices	Heterogenous (all forces)	Homogenous (all forces)	Ribs motion (heterogenous)	Diaph. muscle (heterogenous)
High	20,909	8.13	8.56	8.24	9.68
Medium	16,023	10.41	10.69	10.93	11.76
Low	12,047	12.79	13.07	13.27	14.70

Table 2. Iterations per second of all the simulations on a real patient model (subject 1). The trials have been averaged into a single value.

Resolution	Vertices	Heterogenous (Diaph. muscle)	Homogenous (Diaph. muscle)
High	20,740	6.37	6.84
Medium	10,162	12.05	12.60
Low	5,185	19.39	19.69

Patient-Specific Models

Our method can be used on real patients. To test it, we first took two sets of 4D CT scan data acquired during normal tidal breathing, therefore the diaphragm motions were not as large as in the virtual model described above. We then acquired two MRI data sets from a volunteer subject at two different levels of breath holding.

The diaphragms were manually segmented using ITK-SNAP.[5] A mesh was then generated using the CGAL library[6] with a mesh angle of 5, mesh distance of 1 and mesh radius of 1. The mesh was smoothed using `vtkWindowedSincPolyDataFilter()` with a pass band of 0.2, followed by a 50% decimation, followed by another 200 iterations of windowed sinc smoothing. An example of a segmented and meshed diaphragm is presented in Fig. 4. The mesh is composed of 6633 nodes and 13,267 triangles.

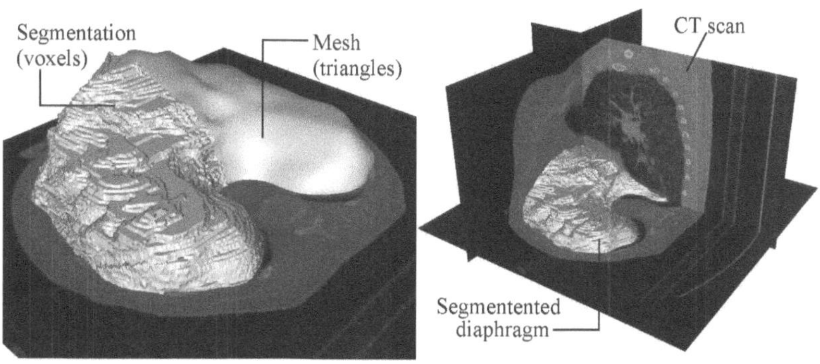

Fig. 4. *Left*: Segmented diaphragm inside CT scan and *right*: combination of segmentation and the mesh.

The rib cage and the lungs were segmented by the levelset method. Rib motion can be neglected in this case because of the normal tidal breathing and the position of the patient lying down on the scanning table. This assumption has been validated by computing the displacement of the ribs. Therefore, only the diaphragm relaxation needs to be taken into account from the initial exhale to the final exhale. The diaphragm will be attached (as in the virtual model) to the locations described in Section 2. The simulation is applied with small forces on the previously defined action line and the lungs are linked to the diaphragm with springs.

The simulation was compared to the imaging data at the same point in the breathing cycle by exporting the resulting lung geometry from the fine-tuned simulation and creating a mesh from the segmented 4D CT diaphragm. The differences were then measured and studied with the MESHDEV[7] software that computes

[5] http://www.itksnap.org/

[6] http://www.cgal.org/

[7] http://www.meshdev.sourceforge.net/

the distance between two triangular meshes. The result can be seen in Fig. 5. The image on the left represents the distance between the surface of the lungs from the CT scan at beginning and end of inhale. The image on the right represents the distance between the simulated end of inhale and the end of inhale extracted from medical images. Finally, Table 2 provides a succinct summary of the computing time. It can be seen that it is slower than the virtual simulator, attributed to the high count of vertices involved in the simulation.

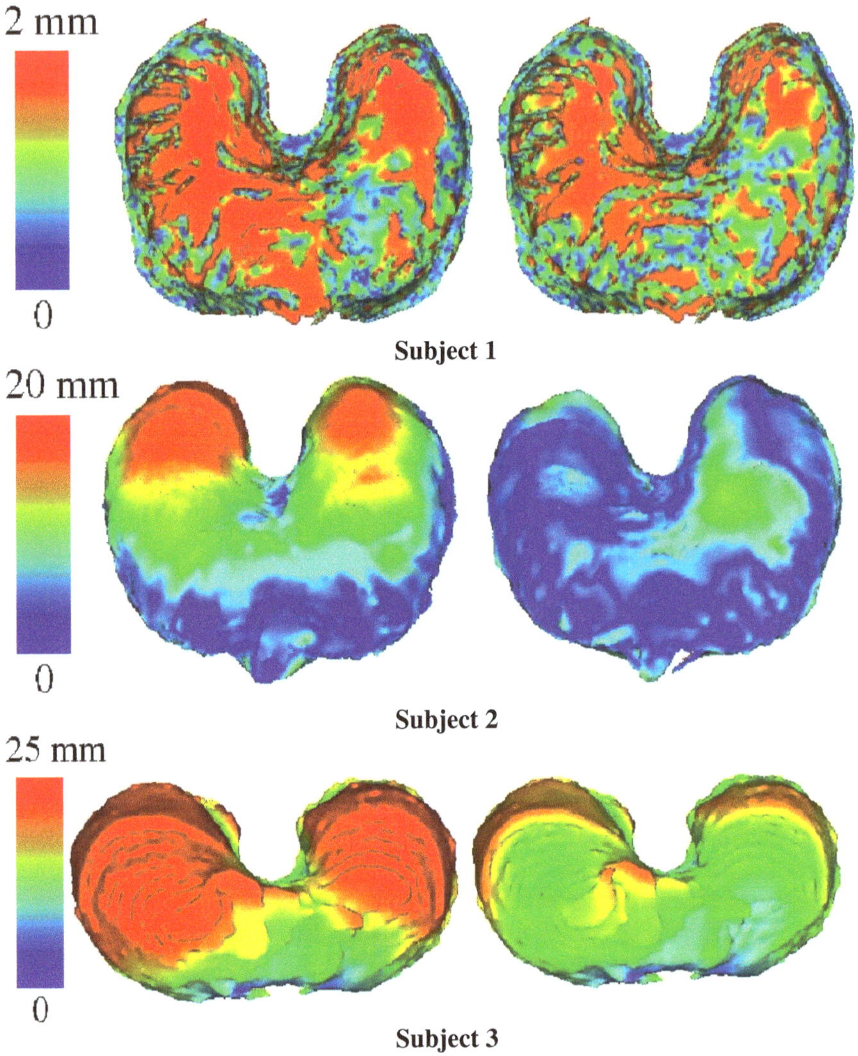

Fig. 5. Distance error measurement on the diaphragm between *left*: beginning and end of real inhale and *right*: simulated end of inhale and real end of inhale.

Discussion

The physiological behaviour of this generic model was discussed with our clinical collaborator, who validated that the antero-posterior, transverse and vertical dimensions of the thorax are increased as illustrated by the model. It was also pointed out that only the domes descend when breathing lightly, whereas with a deep breath, the domes descend further and the central tendon can move from the level of vertebrae T8 to T9. In full exhale, the right dome reaches the 4th intercostal space and the left dome reaches the 5th rib. This complex behaviour further justifies the need for a heterogenous model such as that presented here.

Concerning the patient-specific model, it can be seen that the main output of the simulation is the relaxation of the two domes. The contribution is only slightly significant on subject 1. This is mainly due to the diaphragm displacement (beginning to end of inhale) which is not large enough to correctly appreciate the advantage of the simulation. This displacement is much larger with the 4D CT scan of subject 2 and with the MRI of subject 3. The average diaphragm displacements are, respectively, 5.8 mm with a standard deviation of 5 and 16 mm with a standard deviation of 9.4 mm. The aim of our simulator is to obtain a geometry resulting from the simulation that is as close as possible to the final position of the diaphragm. Figure 5 shows a clear correlation between both meshes, with average displacements of 4.29 mm with a standard deviation of 3.6 mm for subject 2, and 12 mm with a standard deviation of 6.7 mm for subject 3. The diaphragm contraction/relaxation modeling is good enough to simulate a realistic behaviour for liver access procedure training, but it would need to be further improved to be used in radiotherapy where precision is crucial.

The performance evaluations of Tables 1 and 2 show that the addition of the tensegrity constraint makes a slightly slower simulation, but it is still more than adequately fast. The recorded times were very similar. This demonstrates that, even in large models with hundreds of rigid links, the computational time is not massively increased.

Rib kinematics makes a considerable difference to the speed. This is a result of some large equations being solved to update the locations at each iteration. If a patient is observed to only be using the diaphragm and no rib action during respiration (as subjects 1, 2 and 3), a large computational cost can be avoided. Another influencing factor on the speed of a simulation is the size of the mesh. Both virtual and patient models were tested with different resolutions. The evaluation shows an intuitive linear relationship between the number of vertices in the scene and the time taken in computation.

Conclusions

We have presented here a novel methodology to simulate the behaviour of the diaphragm. It is based on a heterogenous model composed of mass–spring and tensegrity elements. These latter elements are used to rigidify the tendinous tissue. The physiological boundary conditions have been carefully studied and applied to our model: the diaphragm motion follows more than one parameter, thus our model also takes into account the influence of the rib kinematics, the muscle natural contraction/relaxation and the motion of the sternum.

Our implementation runs in real time and can be fine-tuned by adjusting the simulation parameters, e.g. if the patient is holding her breath or suddenly having a normal tidal breathing. The anatomy and the physiology can also be customized by generating patient-specific segmentations, mesh models and individual diaphragm actions. In the future, we aim to further validate our method and apply it to liver biopsy planning and rehearsal, as well as to monitor lung motion in a radiotherapy context.

References

1. Shirato H et al. (2006) Speed and amplitude of lung tumor motion precisely detected in four-dimensional setup and in real-time tumor-tracking radiotherapy. International Journal of Radiation Oncology, Biology, Physics 64(4):1229–1235
2. Rohlfing T et al. (2004) Modeling liver motion and deformation during the respiratory cycle using intensity-based free-form registration of gated MR images. Medical Physics 31(3):427–432
3. Siebenthal M, Szekely G, Lomax AJ, Cattin P (2007) Intersubject modelling of liver deformation during radiation therapy. In: MICCAI (1), pp. 659–666
4. Promayon E (1997) Modelling and simulation of the respiration. Ph.D. thesis, University Joseph Fourier of Grenoble
5. Zordan V et al. (2004) Breathe easy: Model and control of simulated respiration for animation. In: ACM SIGGRAPH Symposium on Computer Animation
6. Villard PF, Bourne W, Bello F (2008) Modelling Organ Deformation Using Mass-Springs and Tensional Integrity. In: Proceedings of the 4th international Symposium on Biomedical Simulation, LNCS, vol. 5104, pp. 221–226
7. Ingber D (2000) Opposing views on tensegrity as a structural framework for understanding cell mechanics. Journal of Applied Physiology 89:1663–1678
8. Whitelaw WA (1987) Shape and size of the human diaphragm in vivo. Journal of Applied Physiology 62(1):180–186
9. Gauthier AP, Verbanck S, Estenne M, Segerbarth C, Macklem PT, Paiva M (1994) Three-dimensional reconstruction of the in vivo human diaphragm shape at different lung volumes. Journal of Applied Physiology 76(2):495–506
10. Aladin B, Rodarte J (1994) Inferences on passive diaphragm mechanics from gross anatomy. Journal of Applied Physiology 77(5):2065–2070
11. Meier M (2005) Real-time deformable models for surgery simulation: A survey. Computer Methods and Programs in Biomedicine 77(3):183–197
12. Nedel LP, Thalmann D (1998) Real time muscle deformations using mass-spring systems. In: CGI, pp. 156–165

13. Didier AL, Villard PF, Bayle JY, Beuve M, Shariat B (2007) Breathing Thorax Simulation based on Pleura Behaviour and Rib Kinematics. In: IEEE Information Visualisation – MediVis, pp. 35–40

Toward Anatomical Simulation for Breath Training in Mind/Body Medicine

Benjamin Sanders[1], Paul DiLorenzo[2], Victor Zordan[3], and Donald Bakal[4]

Abstract The use of breath in healing is poorly understood by patients and professionals alike. Dysfunctional breathing is a characteristic of many unexplained symptoms and mind/body medical professionals seek methods for breath training to alleviate such problems. Our approach is to re-purpose and evolve a recently developed anatomically inspired respiration simulation which was created for synthesizing motion in entertainment for the use of visualization in breath training. In mind/body medicine, problems are often created from patients being advised to breathe according to some standard based on pace or volume. However, a breathing pattern that is comfortable and effortless for one person may not have the same benefits for the next person. The breathing rhythm which is most effortless for each person needs to be dynamically identified. To this end, in this chapter, we employ optimization to modify a generic model of respiration to fit the breath patterns of specific individuals. In practice, the corresponding visualization which is specific to individual patients could be used to train proper breath behavior, both by showing specific (abnormal) practice and recommended modification(s).

Introduction

The breath is the ultimate healing heuristic in mind/body medicine; yet the use of the breath in healing is poorly understood by patients, professionals, and the public at large. In mind/body medicine, effortless breathing is a heuristic to guide patient development of somatic awareness in the management of medically

[1] University of California, Riverside, CA, USA
bsanders@cs.ucr.edu

[2] The University of California, Riverside, CA, USA
Dreamworks Animation SKG, Glendale, USA
pdiloren@cs.ucr.edu

[3] University of California, Riverside, CA, USA
vbz@cs.ucr.edu

[4] University of Calgary, 2500 University Drive, Calgary, Canada
dbakal@ucalgary.ca

N. Magnenat-Thalmann et al. (eds.), *Recent Advances in the 3D Physiological Human*,
DOI 10.1007/978-1-84882-565-9_7,

unexplained symptoms (chronic fatigue, fibromyalgia, irritable bowel, non-cardiac chest pain). Dysfunctional breathing is a characteristic of the majority of cases related to these symptoms and therefore careful breath training is both critical but difficult to administer, in part because breathing is a highly individualized attribute that is under both conscious and unconscious control.

Our approach is to develop motion simulation of respiration which will be used for the purposes of educating and training patients. Visualization of the respiratory system is made possible through an anatomically inspired, physically based model of the torso. This model includes all of the key muscles, bones, and connective components found in the human trunk. We have developed this simulation for use in the synthesis of realistic breath motion for visualization of healthy and paradoxical breathing. A similar respiration system [1] is currently in place used as a tool for educating individuals about breath mechanics for stress management by trainers of Blue Cross but, in this effort, we extend the simulation with the goal of breath abnormalities heading toward the use of the simulation for breath training. In addition, a second goal is to modify the simulation to be most appropriate and realistic for specific patient breath training.

Motivating personalized models of breath respiration, in mind/body medicine, problems are often created from patients being advised to breathe according to standard metrics, for example, based on pace or volume. In practice, a breathing pattern that is comfortable and effortless for one person may not have the same benefits for the next person. The breathing rhythm which is most effortless for a single individual is derived from several factors, including quantifiable physical characteristics such as their stature, height, weight, and so on as well as from immeasurable quantities such as muscle features, voluntary habits that have become subconscious, and potentially injuries and the corrections/allowances individuals have made for them over time. A breath signature needs to be dynamically identified for each person. To this end, we are proposing to explore adaptation methods that employ optimization techniques to modify a generic model of respiration to fit the breath patterns and characteristics of specific individuals. To accomplish this, patients would be recorded using state-of-the-art optical-infrared motion capture equipment employing non-invasive motion markers at key landmarks on the body. The patient's own motion data will be used to derive motion control parameters. From this effort, a visualization that is specific to the patient will be used to train breathing for the individual, both by showing their common (abnormal) practice and recommended modifications.

Related Work

Visual and physical simulation of synthetic anatomical muscles has been described for several applications related to modeling and animation, for example, in the head, neck, and face [2–8], the hand [9–11], and for skeletal muscles [12–17]. Visual simulation of skeletal muscles has been approached procedurally through heuristic shape changes made in response to bone movement [13, 14, 18]. These examples model the change in shape of a muscle through geometric muscle bellies that stretch and deform based on length.

Physically based approaches for skeletal muscles include the early work of Chen and Zeltzer [12] who use a biologically based muscle model to generate proper muscle force and Teran and colleagues who use a finite volume method (FVM) to create a continuous internal tension-based muscle simulation focusing on the muscles of the arm [16, 17]. Both show results of deformation on the muscles systems of a single limb. In addition, Nedel and Thalmann propose the use of a spring–mass system as an alternative for real-time applications [15].

We have found no examples of work which are performed on active muscle deformation for animation of the human body, still fewer examples of the nature and complexity of those described here. Active deforming muscles yield the impetus for respiration physiology, namely the contraction of the diaphragm. Close to our efforts are the models of Kaye et al. [19] who animate deformable lungs for clinical applications based on CT scans and simplified cardiopulmonary mechanics and the constraint-based solver of Promayon et al. [20] which models the deformation of the abdomen during calm breath. However, to the best of our knowledge, ours is the first work to investigate the animation of the torso by simulating the motion of both bone such as the ribcage and deformable subcomponents such as the abdomen.

Motivation: Mind/Body Medicine

Evidence demonstrates that stress may exacerbate or cause illness and that mind/body interventions can improve clinical outcomes. Mind/body medicine is a scientifically validated medical discipline. Interventions, including cognitive/behavioral therapy, relaxation response training, lifestyle modifications addressing nutrition and exercise, and coping strategies, have been applied successfully to decrease the frequency and intensity of medical symptoms and to improve disease management in patients with hypertension, heart disease, chronic pain, insomnia, infertility, menopause, and a host of other conditions with a stress component.

Medically unexplained symptoms account for over half of all outpatient visits with 20–25% remaining chronic and recurrent. Non-cardiac chest pain is estimated to account for up to 70% of all patient presentations with chest pain. With population growth and increasing case load pressure on medical specialists, it is clear that an improved method of patient management is required. Conventional pharmacological, psychological, and alternative treatments (supplements) for medically unexplained symptoms are generally ineffective. Through training in recognition and practice of effortless breathing, patients and physicians will have the means to enhance tacit holistic knowing and self-management through the use of their bodies.

In addition, capturing "dysfunctional," "normal," and "effortless" breathing patterns is not readily achieved with current methodologies. The quantification of breathing is generally achieved through the use of strain gauge devices affixed to the abdomen and chest. These measures provide average data points which are used to make comparisons in terms of static volume differences and rate. These measures do not lend themselves to identifying the subtle motion that captures the actual movement of the upper chest, shoulders, thoracic and abdominal regions during inhalation and exhalation.

Approach Overview

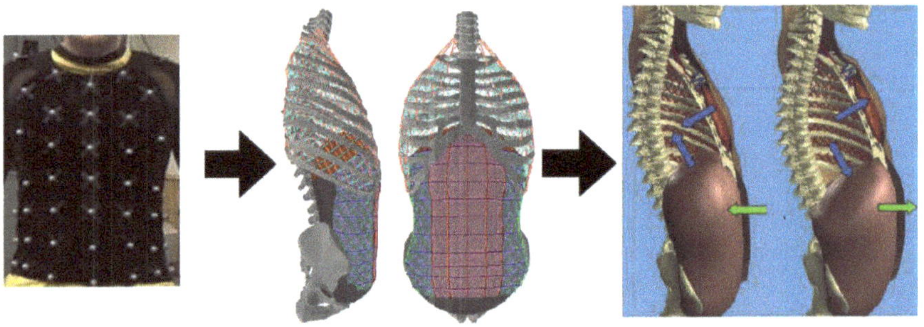

Fig. 1. *Left*: Marker data on the human torso is input. *Center*: A composite trunk simulation for breathing simulation, articulated rigid-body bones and deformable surfaces for the diaphragm and abdomen. *Right*: Through optimization, motion is synthesized for the trunk which matches the motion data.

We propose to use forward dynamic simulation in order to manage the inherent complexity of the processes as well as to increase the benefit of these efforts in fields beyond computer graphics – the dynamics lead to approximations of forces, accelerations, stresses, and other important factors that may be of interest to medical researchers. In computer animation, physical simulation has been used extensively to model rigid-body motion and deformation of bodies (separately), but our choice of modeling bodily function diverges from previous efforts because human physiology readily mixes rigid and flexible components requiring an inherently complex "mixed" simulation structure. We introduced the premiere physical model for animating the movement of human respiration in 2004 [1] and here, we offer extensions that will follow from this initial work leading to more realistic physically based anatomical modeling aimed at breath training for patients.

Further, in medicine, extensive study has gone into very specific anatomical simulations particularly of the heart (hundreds are performing research in cardiac mechanics today) while other biological systems of importance and interest, such as respiration, remain practically untouched. One possible reason for this is the complicated nature of the process. While the heart is largely a "stand-alone" muscle, the respiration system is constructed of rigid bone and deformable muscle, cartilage, and tissue. All of these components must work together by actively contracting and passively responding appropriately in order to form breath. We plan to extend our investigations by creating physiologically valid, mixed rigid-body/deformable models and this chapter represents a step in the process. Here, within the scope of breath training, we excite the system using motion capture to be able to encode individual breath signatures onto the model (see Fig. 1).

Torso Simulation

Our torso simulation is broken down into five main components: spine, ribcage, diaphragm, abdominal cavity, and clavicles – all are highlighted in Figs. 2–4. We use a mixed combination of torque actuators, applied to the spine, and a Hill-type muscle model for muscle elements that we use in the ribcage, diaphragm, abdomen, and clavicles. The torso simulation contains 40 rigid bodies and a deformable body for the abdominal cavity, 1100 Hill muscle elements, and 72 degrees of freedom in the spine. We followed our previous techniques described in [1] with a few notable exceptions, for example, previously we allowed no articulation in the spine and treated it as a single rigid body.

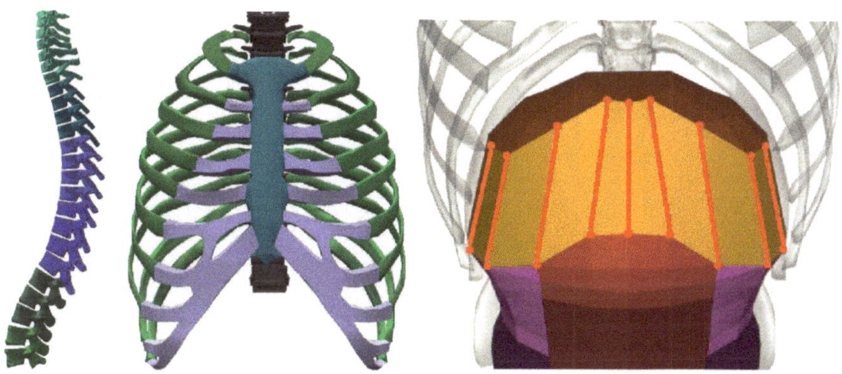

Fig. 2. *Left:* The torso simulation spine colored to denote sections. *Center*: The ribcage and sternum, also denoted by color. *Right*: Bright red line segments highlight the contractile muscles of the diaphragm.

Spine. The actual human spine is composed of 24 vertebrae in 3 separate sections: 5 in the lumbar, 12 in the thoracic, and 7 in the cervical. Laughter often includes noticeable movement in the spine due to the violent, involuntary contractions of the abdomen [21]. After a number of empirical tests conducted in our study of laughter, we concluded that combining groups of vertebrae gives the articulation we desire along with reasonable stability at a manageable timestep. In particular, as seen in the coloring of Fig. 2, we break up the lumbar into two sections (L1–L3 and L4–L5), the thoracic into three sections (T1–T4, T5–T8, T9–T12), and the cervical into two sections (C1–C3, C4–C7). In addition, the skull is added as a separate rigid body connected to the C1–C3 section. The spine is rooted at the pelvic girdle and each joint is treated as a ball joint. Back stability muscles are accounted for by proportional-derivative (PD) servos.

Ribcage. The ribs are each rigid bodies attached to the spine via ball joints. The sternum and cartilage complete the ribcage as shown in Fig. 2. We treat the portions of the cartilage as rigid, rather than approximating them as a more realistic, very stiff but deformable material. The reasoning for this is largely practical – to avoid the stiff deformation which would immensely slow down the computation time of the simulation. We use universal joints to attach the cartilage to the sternum to avoid creases in those joints. Unlike previously, we add degrees of freedom in the connections between the sternum and cartilage so that we can generate a much wider range of motion for the ribcage.

Diaphragm. Our diaphragm is quite novel in contrast to previous approaches because we explicitly model the central tendon (shown in rust color at the top of the diaphragm in Fig. 2). The diaphragm, which is shaped like an inverted bowl, contracts largely along the "sides" of the bowl while the central tendon at its base sees much less (passive) stretching or contraction. Our diaphragm is modeled so that the muscle elements are aligned in the direction of contraction making the diaphragm much more effective as a "pump" for the air in the lungs.

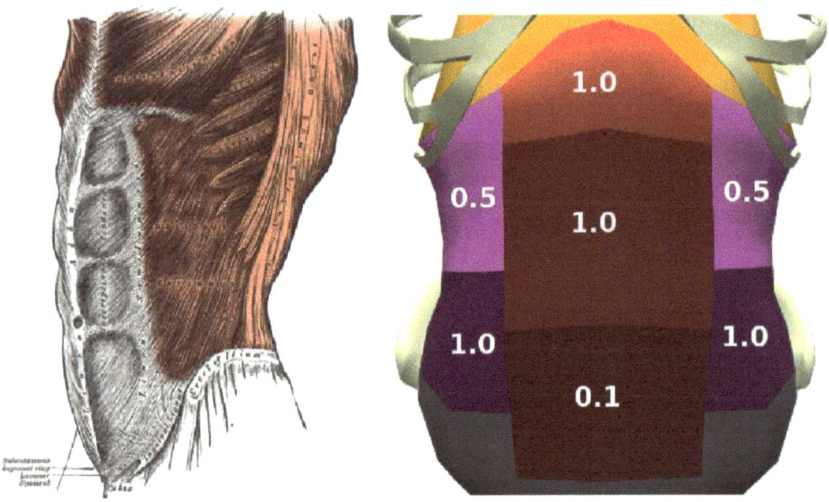

Fig. 3. *Left*: The obliques externus abdominis [22]. *Right*: The proportions shown are ratios of activation for the colored regions of the abdomen adapted from [23].

Abdominal Cavity. Our deformable abdominal cavity follows our previous design and methods by including a volume preservation method. In many forms of paradoxical breathing, activation of the abdomen is very important. According to [23], the pattern of activation of the abdomen is not uniform. We use their findings to identify five regions in the abdomen (shown in multiple colors in Fig. 3) that we activate proportionately, as shown, based on their findings.

Clavicle. In many stress-related cases, shoulder breathing is prevalent. In our experiments, we include movement of the clavicle, which effects the shoulders and subsequently, the arms. Our clavicle is attached at the front of the ribcage close to the top of the sternum using a ball joint. To capture the interplay of the clavicle, the ribcage, and the spine we attach a fairly extensive set of muscles that have proper origin and insertions between these subsystems (see Fig. 4). These muscles propagate activity between the components. In addition, we can use their contraction to control and add additional movement to the clavicle.

In many stress-related cases, shoulder breathing is prevalent. Therefore, in our refined torso simulation, we include improved movement of the abdomen and clavicle, which effects the shoulders and subsequently, the arms.

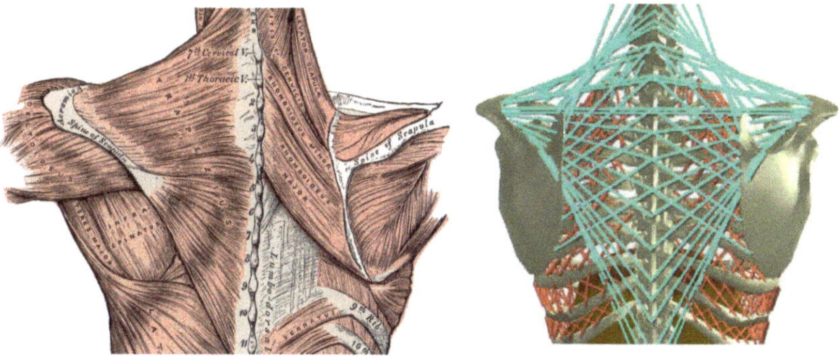

Fig. 4. *Left*: Anatomical shoulder muscles [22]. *Right*: Torso simulation shoulder muscles.

Control

To control the torso simulation, we propose a three-tier hierarchical controller. At the lowest level we compute forces using Hill muscle elements for the respiratory muscles, the abdomen, and the muscles attached to the clavicle. At the mid-level, we group like muscle components and activate them in unison. For example, all the muscle elements in the diaphragm receive the same inputs, even though each is applying a different force as appropriate to its specific conditions. At the highest level, we generate control signals based on the high-level characteristics of the motion. We apply force through each of the muscle elements using a Hill muscle model. Our formulation is derived from insights drawn from previous research on Hill-type muscle modeling, particularly [24]. The basic structure of the muscle model includes a passive element which contributes forces when the muscle is stretched and the contractile element which contributes force when the muscle is activated. For details regarding our Hill muscle model, we refer interested readers to see [21]. PD servo control of the spine allows additional control for full-torso bending as seen in Fig. 5.

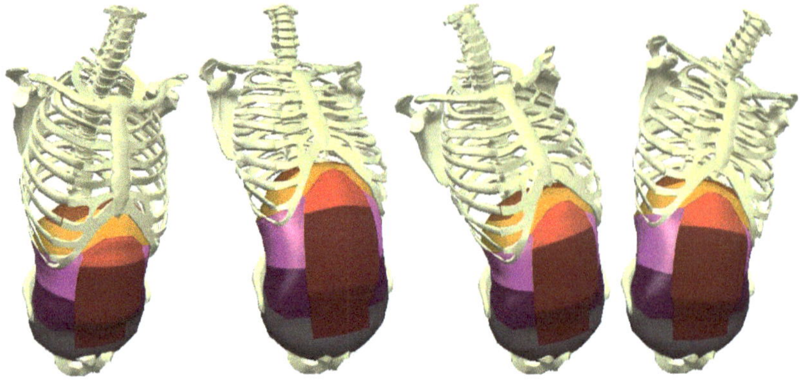

Fig. 5. Flexion, extension, and lateral bending (left and right) of the spine.

Functionally, the control that drives breath is generally split between the two moving subsystems, the intercostals and the diaphragm. These parts move in a synchronized manner and do affect each other but have unique control input based on their own neural activations. This split is reflected in the control of the human respiration system with the nerves which supply contraction input to the muscle groups stemming from different portions of the spine [25]. Further, during effortless breathing, only the diaphragm may be activated, without any intercostal contraction. We control activation for muscle groups by changing the contraction level over time for the muscle elements in the intercostals and the diaphragm. In Fig. 6, we show the result of simple periodic contraction levels for activation that are able to synthesize breath styles with quantitative similarities to actual human breath results.

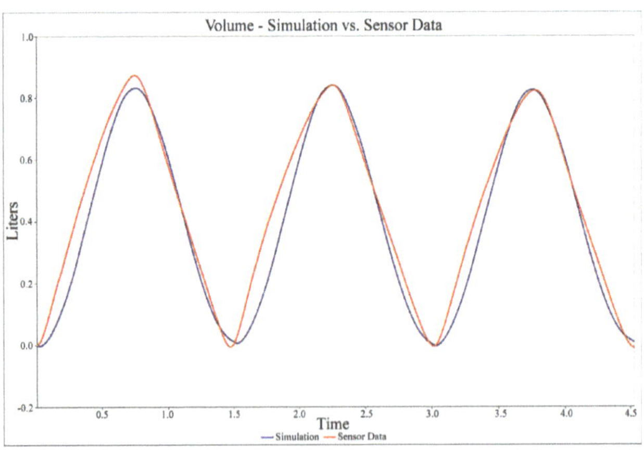

Fig. 6. Tidal volume comparison between simulation and air-flow sensor from a human subject.

Driving with Human Motion

Motion capture data, recorded measurements taken from human subjects, faithfully reproduces many movements of the human body. We use a marker-based Vicon motion capture system to measure chest and stomach deformations using an ample suite of markers. In addition, the placement of these markers corresponds to anatomical landmarks such as the sternum, clavicle, and navel. The system is capable to track reflective markers attached to the body with a 3D accuracy within 1 mm.

Of course, the motion recorded only captures the gross external effects of the body. Seeing internal effects, such as the diaphragm, is not possible without inference. For the purposes of visualizing anatomy for humans, captured data are not valuable because it is not possible to record the movement of the internal organs and muscles directly. In contrast, simulation holds the promise of realism and generality because it is based on the physical laws that govern the real world. However, motion created through simulation requires control in order to perform coordinated behaviors realistically. Our assumption is that given the breath motion data, i.e., only 3D points, we can generate a visualization that is the integration of the breath simulation with the marker data using the data to determine control for specific recordings.

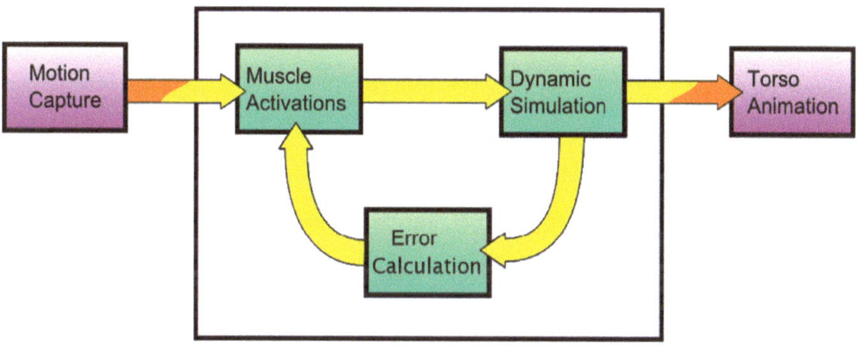

Fig. 7. An overview of the optimization process.

Optimization. Motion data cannot be directly used to control the simulation. However, we can determine the activation levels required to achieve the desired motion through optimization. We chose to activate the main muscle groups that contribute to respiration: the outer intercostals and diaphragm which are used for inhalation and the inner intercostals and abdomen which are used for exhalation. Although the abdomen is broken up into sections, we determine a single value for their collective activation based on the findings in the literature [23] and use it to compute the specific activation levels of each section as described. To further simplify, we link the timing parameters of the outer intercostals with the diaphragm and the inner intercostals with the abdomen. For the four different muscle groups, we only allow the optimizer to choose the maximum muscle activation for each muscle group. Thus, the optimizer has only four parameters to optimize over. Despite having a small number of parameters, the three-tiered hierarchy in place serves as the basis for the visually rich, complex motions produced from our system when acting upon the optimized parameters. We experimented with activation inputs such as sinusoids, step function, and linear interpolation of control points, along with different parameter controls such as phase and period. We found that piecewise-linear sawtooth inputs produced the most visually pleasing results and comes closest to the input data of different types of subjects.

Rather than optimizing for an entire breath sequence simultaneously, we use an idealized (average) cycle of breath from each patient. This "breath signature" is derived by analyzing characteristics of motion data. This trade-off allows computation of a small set of parameters in exchange for more specific optimizations applied to each breath cycle. It also allows for motion-cycling when the patient views her/his own breath. By placing virtual landmarks on the simulation, the system "records" virtual motion capture data (from the simulation), thereby creating a one-to-one comparison of the simulation and the human. We use OPT++'s [26] parallel direct search method, which is an optimization technique that does not use derivative information to solve for the parameters. We used this method to take advantage of the parallelism in OPT++ since the function evaluation takes a considerable amount of time due to the simulation time of the torso system. Our cost function is defined as the error between the human and simulated motion data.

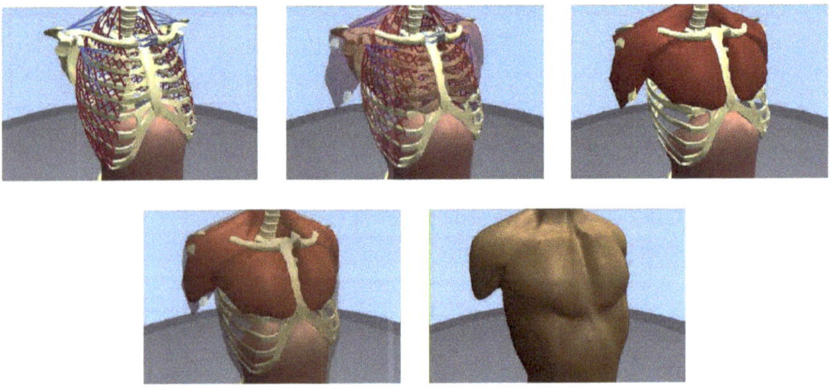

Fig. 8. Layers of the torso model used for demonstrating breath simulation.

The result is a motion synthesis that replicates the patient's own motion. The motion can be viewed transparently, so that the patient and doctor can observe the internal function. Also, the relationship between the components can be seen. An example of the variety of visualizations possible appears in Fig. 8. We expect that doctors will be able to control the transparency of the layers based on the demands of the situation.

Results and Analysis

We recorded the breathing behavior of four subjects of varying height and weight using a motion capture system. After recording the subjects, we determine their breath signature. The marker locations from the motion capture system are matched with our torso simulation. Then the optimizer iteratively tests a number of samples of breath control parameter sets based on the specified constraints. The system calculates the average displacement error, $error_p$, and velocity error, $error_v$, between the motion capture markers and each synthesized run of the torso simulation. We define

$$error_p = \frac{\sum_{0<i<m} \sum_{0<j<n} d(i,j)}{n * m} \tag{1}$$

where m is the number of frames, n is the number of motion capture points, $d(i,j)$ is the distance between the motion capture marker and the torso simulation marker. We define

$$error_v = \frac{\sum_{0<i<m} \sum_{0<j<n} v(i,j)}{n * m} \tag{2}$$

where $v(i,j)$ is the instantaneous velocity difference. Table 1 shows the results of the best solution found by the optimizer for each subject as well as the activation chosen by the optimizer. An interpretation of the values of the subjects allows us to make high-level assessments regarding individual breathing style based on the variability in activation between the subjects. Subject 1 primarily uses the ribcage with slight use of the diaphragm during inhale and a small amount of activation of both the ribcage and abdomen in exhalation. Subject 2 only uses the ribcage to inhale and no muscle activation in the diaphragm. For exhalation, this subject passively exhales. In contrast, subjects 3 and 4 use an even mix of ribcage and diaphragm activation to inhale, while only subject 3 uses muscle activation to exhale. Our system was able to capture these different types of breathing with an average displacement error of less than 7 mm per marker and an average velocity error around 6 mm/s (see Fig. 9).

Paradoxical Breathing. A common paradoxical breath pattern takes the form of contraction of the abdomen during inhale. We use our system to analyze this dysfunction. We compared normal breathing with a simulation of this type of paradoxical breathing using tidal volume. We construct a paradoxical breathing example which was verified by our breathing expert, Dr. Bakal (co-author). We compare this simulation to the same example without the activation of the abdomen. The results of our experiment can be seen in Figs. 9 and 10. We can see that during paradoxical breathing, the simulation determined the tidal volume to be approximately 400 ml. If we remove the abdomen activation, the simulation determined the tidal volume to be approximately 800 ml. Clearly, there is a loss in tidal volume when someone breathes in this paradoxical manner compared to normal breathing. Figure 9 (*right*) shows a few comparison frames between the normal breathing simulation (*top*) and the paradoxical breathing simulation (*below*). The benefit of contracting the diaphragm is absent during the paradoxical breathing while it is used prominently in normal breathing. This observation argues for the use of a simulation such as our model since the effects of the diaphragm movement cannot be seen from the exterior of the trunk.

Table 1. Table showing the average positional and velocity error between the motion capture markers and the simulation markers for the best optimized run for each subject. The activation levels for the outer intercostals, diaphragm, inner intercostals, and abdomen are shown.

	$error_p$ (mm)	$error_v$ (mm/s)	Outer	Diaphragm	Inner	Abdomen
Subject 1	5.66	4.64	0.305	0.040	0.018	0.007
Subject 2	8.40	9.29	0.488	0.002	0.002	0.002
Subject 3	8.62	6.17	0.254	0.099	0.033	0.010
Subject 4	4.48	4.11	0.261	0.106	0.000	0.000
Average	6.79	6.05				

In our next step, for future work, we plan to begin to look into the recording of a patient's exact body to match physique. Techniques that allow this have already been reported in computer graphics [27]. In addition, it is critical that we investigate methods for validation of our model so that it can be viewed, especially by medical personnel, with more confidence. Currently, failure for the optimization is possible but could act as an indicator of an unusual circumstance. One viewpoint is that rather than being construed as a failed optimization, a poor fit between our model and a subject's data may be an indicator of an anomaly – that something out of the ordinary is taking place and/or the patient's breath is truly exceptional. In such a case, further testing could be prescribed, including MRI or the like.

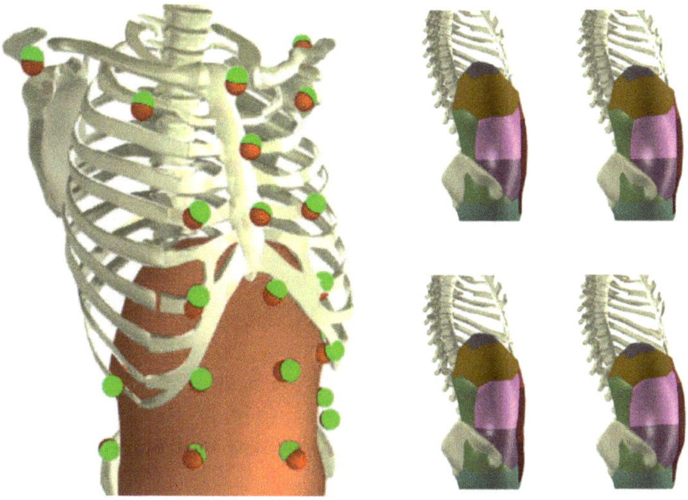

Fig. 9. *Left*: Comparison of the motion capture and simulated markers overlaid on the torso simulation. *Right*: Diaphragm comparison between the simulated paradoxical breathing (*top*) and normal breathing (*bottom*).

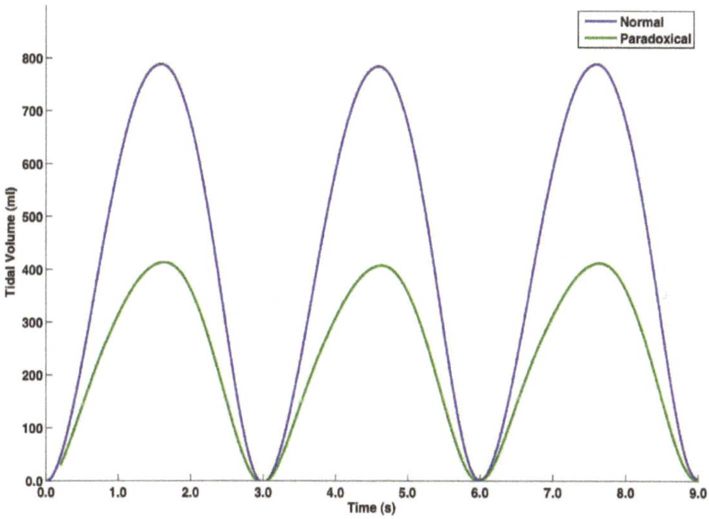

Fig. 10. Tidal volume comparison between normal and paradoxical breathing.

Conclusions

Our research is still preliminary. However, even if we see a component of failure in being able to personalize our respiration model (i.e., the riskiest thrust of the research) we expect the respiratory modeling will be of grand benefit. In addition to the novel and important testbed formed by breath training, simulation of respiration required a substantive understanding of the modeling. That said, we have made several tests which reveal that our model is indeed quite realistic. Quantitatively, the tidal volumes and pressure results of our models are already quite close to the real human system. This is not surprising given the care and attention we placed on our selection of parameters for our system: our model matches the anatomy of respiration very closely. For example, upon cursory inspection, the skeletal anatomy is close to the real human in size, muscle insertion and origins are accurate, and our muscle activation using the Hill-type muscle model is very progressive by any standard. Including more aspects and input from respiration experts will undoubtedly improve this trend.

In addition, based on anecdotal evidence, we already anticipate an easy acceptance of the generic respiration system as a tool for visualizing and training breath. Qualitatively, breathing *with* the system is "natural," i.e., synchronizing breath with the animation has been an unsolicited trend among observers. Several viewers upon their first time viewing of the animations begin to align their breath with the animated breath without any suggestion or indications offered to do so. That is, they involuntarily begin to *"breathe with"* the simulated motion. This is an exceptional phenomenon because it is exactly what is needed for breath training. If the motion capture matching does not reach fruition, we will develop pure simulations and based on our experience, we believe that patients will readily match breathing, especially if encouraged to do so.

References

1. Zordan VB, Celly B, Chiu B, DiLorenzo PC (2004) Breathe easy: Model and control of simulated respiration for animation. In: ACM SIGGRAPH/Eurographics Symposium on Computer Animation, pp. 29–37
2. Lee S, Terzopoulos D (2006) Heads up!: Biomechanical modeling and neuromuscular control of the neck. ACM Transactions on Graphics 25(3):1188–1198
3. Platt S, Badler N (1981) Animating facial expressions. SIGGRAPH Computer Graphics 15(3):245–252
4. Waters K (1987) A muscle model for animating three-dimensional facial expressions. In: Proceedings of SIGGRAPH 87, pp. 17–24
5. Terzopoulos D, Waters K (1990) Physically-based facial modelling, analysis, and animation. Journal of Visualization and Computer Animation 1(2):73–80
6. Lee Y, Terzopoulos D, Walters K (1995) Realistic modeling for facial animation. In: Proceedings of SIGGRAPH '95, pp. 55–62

7. Kähler K, Haber J, Yamauchi H, Seidel H (2002) Head shop: Generating animated head models with anatomical structure. In: Eurographics/ACM SIGGRAPH Symposium on Computer Animation, pp. 55–64
8. Sifakis E, Neverov I, Fedkiw R (2005) Automatic determination of facial muscle activations from sparse motion capture marker data. ACM Transactions on Graphics, 24(3):417–425
9. Gourret J, Magnenat-Thalmann N, Thalmann D (1989) Simulation of object and human skin deformations in a grasping task. In: Proceedings of SIGGRAPH 89, pp. 21–30
10. Albrecht I, Haber J, Seidel H (2003) Construction and animation of anatomically based human hand models. In: Eurographics/ACM SIGGRAPH Symposium on Computer Animation, pp. 98–109
11. Tsang W, Singh K, Fiume E (2005) Helping hand: An anatomically accurate inverse dynamics solution for unconstrained hand motion. In: 2005 ACM SIGGRAPH/Eurographics Symposium on Computer Animation, pp. 319–238
12. Chen D, Zeltzer D (1992) Pump it up: Computer animation of a biomechanically based model of muscle using the finite element method. In: Proceedings of SIGGRAPH 92, pp. 89–98
13. Scheepers F, Parent R, Carlson W, May S (1997) Anatomy-based modeling of the human musculature. In: Proceedings of SIGGRAPH 1997, pp. 163–172
14. Wilhelms J, Gelder A (1997) Anatomically based modeling. In: Proceedings of SIGGRAPH 1997, pp. 173–180
15. Nedel L, Thalmann D (1998) Real time muscle deformations using mass-spring systems. In: Computer Graphics International, pp. 156–165
16. Teran J, Blemker S, Ng-Thow-Hing V, Fedkiw R (2003) Finite volume methods for the simulation of skeletal muscle. In: Eurographics/ACM SIGGRAPH Symposium on Computer Animation, pp. 68–74
17. Teran J, Sifakis E, Blemker S, Ng-Thow-Hing V, Lau C, Fedkiw R (2005) Creating and simulating skeletal muscle from the visible human data set. IEEE Transactions on Visualization and Computer Graphics 11(3):317–328
18. Pratscher M, Coleman P, Laszlo J, Singh K (2005) Outside-in anatomy based character rigging. In: 2005 ACM SIGGRAPH/Eurographics Symposium on Computer Animation, pp. 329–338
19. Kaye J, Metaxas D, Primiano Jr F (1997) A 3d virtual environment for modeling mechanical cardiopulmonary interactions. In: CVRMed, pp. 389–398
20. Promayon E, Baconnier P, Puech C (1997) Physically-based model for simulating the human trunk respiration movements. In: Lecture Notes in Computer Science, Springer Verlag, CVRMED II-MRCAS III First Joint Conference, vol. 1205, pp. 379–388
21. DiLorenzo P, Zordan V, Sanders B (2008) Laughing out loud: Control for modeling anatomically inspired laughter using audio. ACM Transactions on Graphics, SIGGRAPH Asia, 27(5):125:1–125:8
22. Gray H (1918) Anatomy of the Human Body. Lea & Febiger, Philadelphia, Pennsylvania
23. Hoit J, Plassman B, Lansing R, Hixon T (1988) Abdominal muscle activity during speech production. Journal of Applied Physiology, 65(6):2656–2664
24. Zajac F (1989) Muscle and tendon: Properties, models, scaling, and application to biomechanics and motor control. In: CRC Critical Reviews in Biomedical Engineering, vol. 17, pp. 359–411
25. Mines AH (1993) Respiratory Physiology, Raven Press, New York
26. Meza J, Oliva R, Hough P, Williams P (2007) OPT++: An object oriented toolkit for nonlinear optimization. ACM Transactions on Mathematical Software 33(2):Article 12
27. Allen B, Curless B, Popovic (2003) The space of human body shapes: Reconstruction and parameterization from range scans. ACM Transactions on Graphics 22(3):587–594

Simulating the Human Motion Under Functional Electrical Stimulation Using the HuMAnS Toolbox

Martine Eckert[1], Mitsuhiro Hayashibe[2], David Guiraud[3], Pierre-brice Wieber[4], and Philippe Fraisse[5]

Abstract Mathematical models of the skeletal muscle can support the development of neuroprostheses to restore functional movements in individuals with motor deficiencies by means of functional electrical stimulation (FES). Since many years, numerous skeletal muscle models have been proposed to express the relationship between muscle activation and generated force. One of them (Makssoud et al.), integrates the Hill model and the physiological one based on Huxley work allowing the muscle activation under FES. We propose in this chapter an improvement of this model by modifying the activation part. These improvements are highlighted through the HuMAnS (**Hu**manoid **M**otion **An**alysis and **S**imulation) toolbox using a 3D biomechanical model of human named Human 36. This chapter describes this toolbox and the software implementation of the model. Then, we present the results of the simulation.

Introduction

Different approaches to model the electromechanical behavior of skeletal muscles with FES signal as inputs have been presented in previous studies [1, 2]. Thanks to the development of biomechanics, different musculoskeletal human

---

[1] IMERIR, Perpignan, France
martine.eckert@imerir.com

[2] DEMAR Project, INRIA – LIRMM, Montpellier, France
mitsuhiro.hayashibe@lirmm.fr

[3] DEMAR Project, INRIA – LIRMM, Montpellier, France
david.guiraud@lirmm.fr

[4] BIPOP Project, INRIA Grenoble, France
pierre-brice.wieber@inrialpes.fr

[5] DEMAR Project, INRIA – LIRMM, Montpellier, France
fraisse@lirmm.fr

N. Magnenat-Thalmann et al. (eds.), *Recent Advances in the 3D Physiological Human*,
DOI 10.1007/978-1-84882-565-9_8,

models have also been developed and available. The general musculoskeletal model of whole body and its dynamics-computation method were established [6, 7]. The effective computational algorithms were pursued for inverse and forward dynamics on the basis of efficient multibody dynamics computation algorithms [8]. In these works, the Hill-type model [9] is normally used for the calculation of muscle force. In the HuMAnS toolbox (Humanoid Motion Analysis and Simulation toolbox developed at the INRIA Rhône-Alpes [5]), a 3D biomechanical model of the human has been developed: the Human 36 model in which the mathematical muscle model proposed by DEMAR project [1] has been introduced. In this muscle model, Hill representation and physiological model based on the work of Huxley [4] are integrated and designed especially for muscle activation through FES.

Using the HuMAnS toolbox, we have modified the initial model in order to improve the output mechanical response. Input command and the force–length relationship were thus enhanced. Then, we observe the results of the modifications. In the first part of this chapter, we give a description of the HuMAnS toolbox and particularly the Human 36 model. Then, we describe the two models of the skeletal muscle we use, and we present the results and discussion of the simulation.

Materials and Methods

The HuMAnS Toolbox and the Human 36 Model

The HuMAnS toolbox developed at INRIA Rhône-Alpes offers tools for the modeling, the control, and the analysis of humanoid motion, being that of a robot or a human. It is an open source software. It is composed of two parts: a first one which is the model generation and a second one which is the simulation, the control, and the analysis tools. A 3D biomechanical model of human has been developed on this software: the Human 36 model (Fig. 1).

The Human 36 model includes the geometric, the kinematic, and the dynamic models of the whole human body and also the interaction of the body with the environment [10]. The dynamic model is a Lagrangian one [5]. The interest of using the Human 36 model for static or dynamic studies has been proved in a previous work [11].

The mathematical model of the skeletal muscle proposed by DEMAR project for FES applications has also been implemented in the Human 36 model. Initially, it is the model proposed by El Makssoud et al. [1, 2].

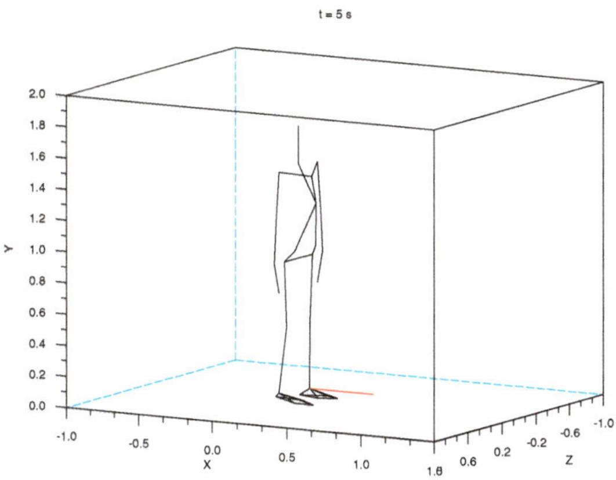

Fig. 1. The Human 36 model.

Skeletal Muscle Models

The muscle models we proposed are composed of two parts of different nature [2]: first, an activation model describing how an electrical stimulus generates an action potential (AP) and initiates the contraction and recruits motor units, then a mechanical model describing the generation of forces and an evolution of lengths [2] (Fig. 2).

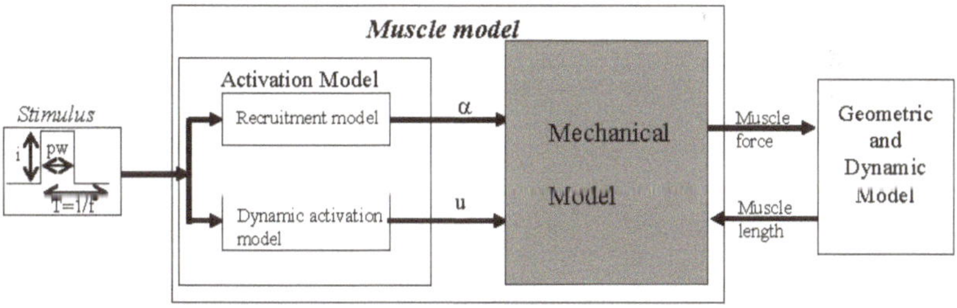

Fig. 2. Electromechanical model of the muscle.

The contraction force generated by a muscle under FES is mainly modulated therefore by the stimulation parameters: intensity (i) and pulse width (pw). For this reason, the muscle model has these stimulation parameters as inputs, the generated force and active stiffness as outputs.

The Skeletal Muscle Model Initially Implemented in Human 36

The command signal $u(t)$ is shown in Fig. 3. It fluctuates between two values U_p and U_m. Its different states correspond to different activations: contraction phase (2), transition phase (1), and relaxation phase (0) (Fig. 3) [1].

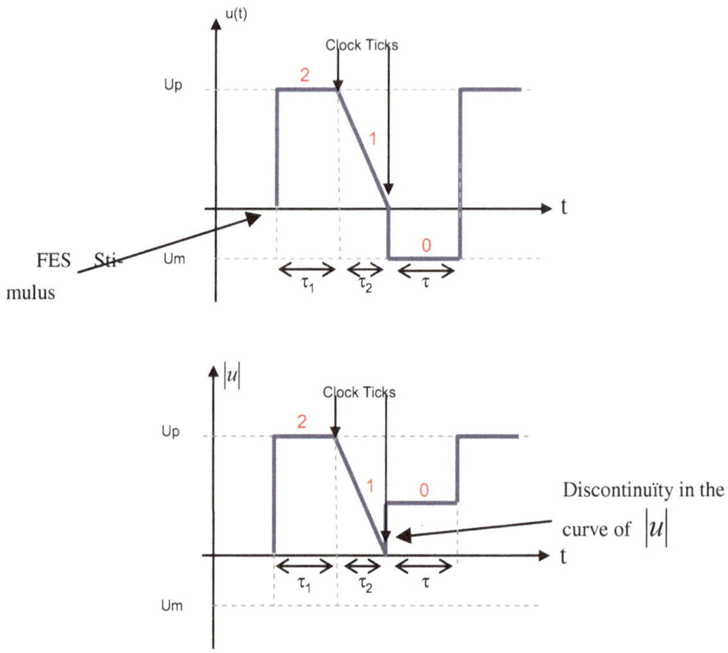

Fig. 3. Command signal [1].

The recruiting rate depends on the width and intensity of the impulsion. Its value is given by [1]

$$\alpha = \frac{1}{2} + \frac{1}{2}\frac{\tanh\left(5\left(\frac{pwi}{pw_{max}i_{max}} - \frac{1}{2}\right)\right)}{\tanh\left(\frac{5}{2}\right)} \tag{1}$$

The mechanical model is composed of a contractile component with the command signal $u(t)$ and α as inputs and passive elements composed of a serial linear spring and a parallel exponential spring (Fig. 4) [1].

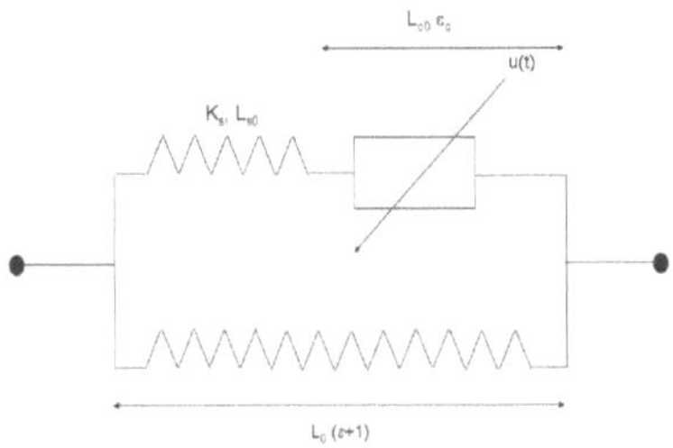

Fig. 4. Mechanical model of the skeletal muscle [1].

The stiffness and the force of a muscle depend on its length:

$$k_0 = k_m\, e^{-\left(\frac{\varepsilon_c}{0.4}\right)^2}$$
$$F_0 = F_m e^{-\left(\frac{\varepsilon_c}{0.4}\right)^2} \tag{2}$$

For the quadriceps and hamstrings muscles, the k_m, F_m, k_s, k_p, k_{ep}, L_{c0} and L_0 values are given in Table 1.

Table 1. Maximum stiffness and force of quadriceps and hamstrings muscles

	Quadriceps	Hamstrings
k_m	10^4 N.m^{-1}	10^4 N m^{-1}
F_m	500 N	500 N
k_s	10^4 N.m^{-1}	10^4 N m^{-1}
k_p	0 N.m^{-1}	0 N m^{-1}
k_{ep}	1 N.m^{-1}	1 N m^{-1}
L_{c0}	8.2 cm	10.7 cm
L_0	49.2 cm	49.2 cm

The dynamic model of contractile component coupled with the linear series spring (Fig. 3) is [1] (where k_c, F_c, ε_c are the state variables)

$$\begin{cases} \dot{k}_c = -(|u| + |\dot{\varepsilon}_c|)\, k_c + \alpha\, k_0 |u|_+ \\ \dot{F}_c = -(|u| + |\dot{\varepsilon}_c|)\, F_c + \alpha\, F_0 |u|_+ + k_c L_{c0}\, \dot{\varepsilon}_c \\ \dot{F}_c = k_s (L_0\, \dot{\varepsilon} - L_{c0}\, \dot{\varepsilon}_c) \end{cases} \tag{3}$$

Then, we can compute $\dot{\varepsilon}_c$ from the command u, the recruiting rate α, the forces F_c and F_0, the stiffness k_c thanks to the second and last equations of the system (Eq. 3):

$$\dot{\varepsilon}_c = \frac{k_s\, L_0\, \dot{\varepsilon} + F_c |u| - \alpha\, F_0\, |u|_+}{k_s L_{c0} + k_c L_{c0} - S_{\dot{\varepsilon}_c} F_c} \quad \text{(with } S_{\dot{\varepsilon}_c} \text{ the sign of } \dot{\varepsilon}_c\text{)} \tag{4}$$

The total force developed by the muscle is

$$F = F_c + \frac{k_p}{k_{ep}} \left(e^{k_{ep}\varepsilon} - 1\right) \tag{5}$$

Modified Activation Model

We propose here an evolution of the DEMAR skeletal muscle model presented above. This modified model has been developed directly from the Huxley [4] and the Hill models [3] and its parameters estimated on animal muscles.

Compared to the previous ones, it includes a new recruitment function and more accurate dynamics, including a force–length relation at the microscopic scale.

A new command signal $u(t)$ is implemented (Fig. 5).

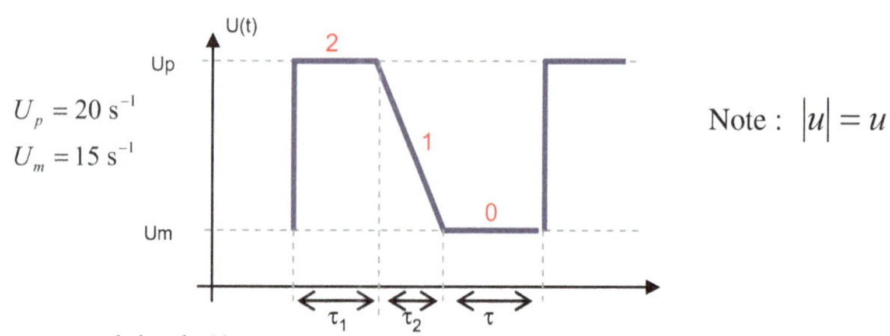

Fig. 5. New command signal $u(t)$.

The contraction–relaxation cycle is considered therefore to stick to the previous simple three phases.

The stronger the intensity (i) is and the longer the pulse width (pw) is, the bigger the contraction of the muscle will be. We propose therefore a new recruitment function with these two parameters partially decoupled:

$$\alpha(p\omega, i) = d(\tanh(R - c) + \tanh(c))$$
$$R = b\left(1 + a_{p\omega}\frac{p\omega}{p\omega_{\max}}\right)\left(1 + a_i\frac{i}{i_{\max}}\right)\left(\frac{q}{q_{\max}}\right) \tag{6}$$

Implementation and Results

The new model presented above has been implemented in the Human 36 model [5]. This part describes the results obtained.

Modification of the Activation Model

We have modified the shape of $u(t)$ and α in the initial model. Only four muscles are modeled: the quadriceps and the hamstrings of the two legs.

We have applied a swing up regular movement to the Human 36 model. It means that at the beginning, the human body is standing and the legs are straight. Then, for the two legs, we stimulate the hamstrings during 30 ms to the bending position after we simulate the quadriceps during 30 ms to the extension position and so on.

The movement is computed for both knees. For the right knee, a free movement is performed whereas for the left knee isometric case is studied.

The results concerning the rotation angle of the right knee (Q) (the left knee does not move), the torques, the active stiffness (k_c), the force (F_c), and ε_c of the contractile element are presented in Figs. 6–10.

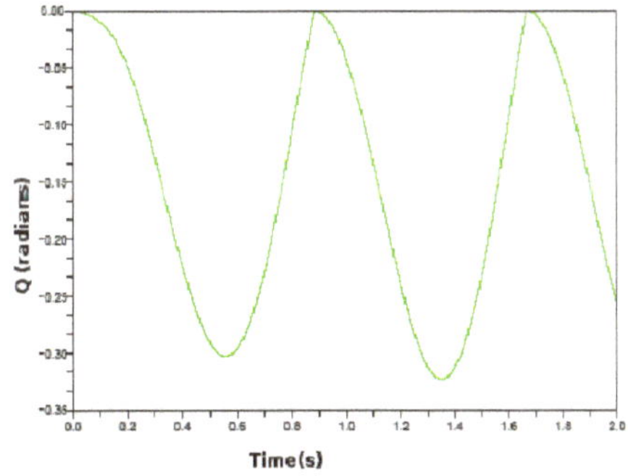

Fig. 6. Evolution of the angle of rotation (Q) when the new activation model is implemented.

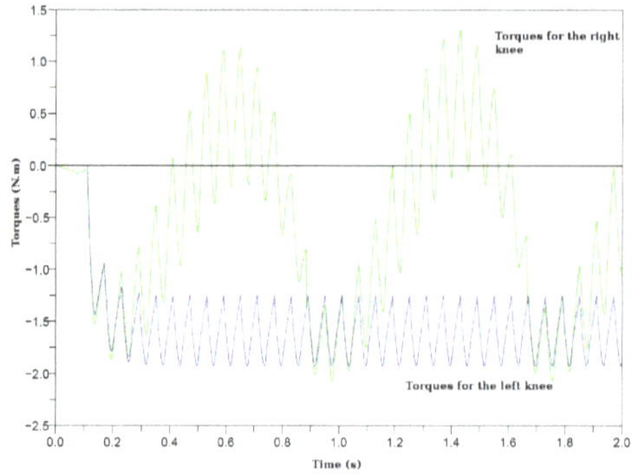

Fig. 7. Evolution of the torques when the new activation model is implemented.

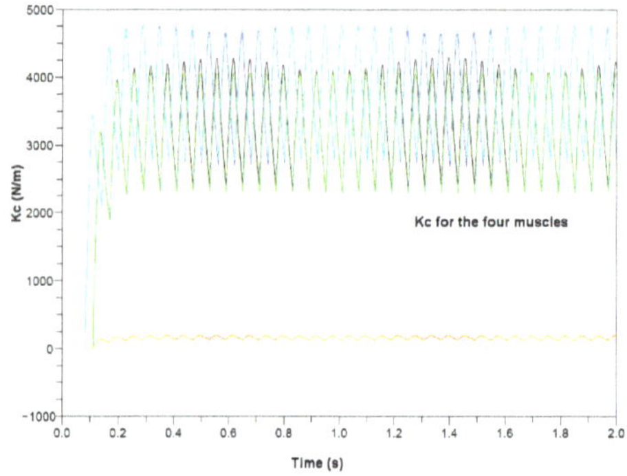

Fig. 8. Evolution of the contractile stiffness (K_c) when the new activation model is implemented.

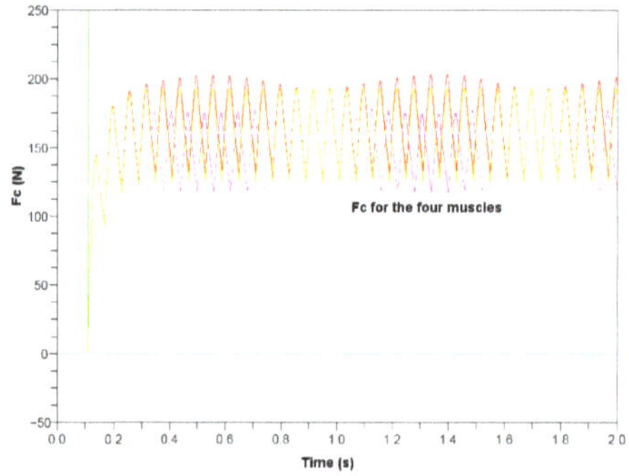

Fig. 9. Evolution of the contractile force (F_c) when the new activation model is implemented.

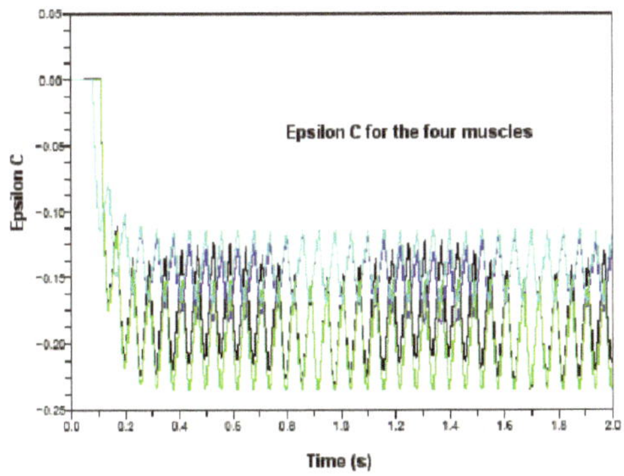

Fig. 10. Evolution of ε_c when the new activation model is implemented.

Table 2. Parameters used in simulation.

Pw	300×10^{-6} s
pw_{max}	$1,4\times10^{-3}$ s
i_{max}	200×10^{-3} A
a_{pw}	–0.428
a_r	1.56
b	2.74
c	0.661
d	0.634
τ_1	20×10^{-3} s
τ_2	10×10^{-3} s

The first model we proposed has been modified due to an insufficient accurate behavior while performing simulation. It was due to two main problems we dealt with: the $|u|$ that was discontinued and the recruitment curve that cannot render the difference of recruitment effect depending on the input used to modulate force (i or pw). We thus change the shape of u without modifying the physiological meaning of this control. For the recruitment, we propose a new recruitment function that takes into account the difference between i and pw modulation. Moreover, in the new model, the recruitment a is kept to 0 when either i or pw is 0 and its cross-sections along the i and pw axes follow sigmoid-like functions as observed experimentally when varying only one of these parameters. Thus both modifications enhance the model with a more realistic behavior, validated on real experiments [12].

With regard to the simulation results we can notice the original following results:

1. The model and its implementation are able to render both isometric and non-isometric contractions; Hill-type models can hardly achieve this with the same set of identified parameters.
2. Relaxation and contraction phase dynamics are different as in the real world, this is also not rendered by linear models of the muscle.
3. Internal variables such as active stiffness, relative length of the contractile element can be observed and simulated; again, classical Hill-type models cannot give such pieces of information.
4. Finally, as contraction and relaxation phase dynamics can be adjusted separately, and owing to the *u* command input, different from the recruitment input, the model is able to predict both twitch and tetanic contractions with a high accuracy. Figures 6–10 show the results obtained, consistent with experimental data (not presented here).

This analysis of the model concerns the software implementation issue, quite complex to achieve, mainly due to the switching and initialization of differential equations and the coupling between muscle's equations and biomechanics state variables such as length. We also qualitatively validate the global behavior of the HuMAnS software with such complex simulation, through the validation of known properties of muscle's behavior listed above.

Conclusion and Perspectives

An initial skeletal muscle model for FES applications was proposed by DEMAR project and has been implemented in the Human 36 model (HuMAnS toolbox [5]). As HuMAnS gives the possibility to test different models of muscle, we simulate here an evolution of the activation model, which is more in accordance with the physiological muscle behavior.

Even though we have already validated this model through experiments on rabbit's muscles, we will now be able to introduce the other skeletal muscles of the human body in order to use the Human 36 model to compare the results given by the simulation and the results obtained with experiments on subjects performing induced FES movements. The model is also able to simulate voluntary contraction but it needs to be modified in order to accept EMG signals as inputs instead of FES parameters. This work is already under development.

Acknowledgments Thanks to C. Azevedo-Coste, M. Benoussaad, and R. Mozul for their precious help in this work.

References

1. El Makssoud H (2005) Modélisation et Identification des Muscles Squelettiques sous Stimulation Electrique Fonctionnelle. PhD Thesis, Université Montpellier 2
2. El Makssoud H, Guiraud D, Poignet P (2004) Mathematical muscle model for functional electrical stimulation control strategies. In: Proceedings of the International Conference on Robotics and Automation (ICRA), pp. 1282–1287
3. Hill AV (1938) The heat of shortening and the dynamic constants in muscle. Proceeding of the Royal Society, London, Series B 126:136–195
4. Huxley AF (1957) Muscle structure and theories of contraction. Progress in Biophysics and Biophysical Chemistry 7:255–318
5. Wieber PB, Billet F, Boissieux L, Pissard-Gibollet R (2006) The HuMAnS toolbox, a homogeneous framework for motion capture, analysis and simulation. In: Ninth International Symposium on the 3D Analysis of Human Movement
6. Delp SL, Loan JP (1995) A graphics-based software system to develop and analyze models of musculoskeletal structures. Computers in Biology and Medicine 25:21 34
7. Delp SL, Anderson FC, Arnold AS, Loan P, Habib A, John CT, Guendelman E, Thelen DG (2007) OpenSim: Open-source software to create and analyse dynamic simulations of movement. IEEE Transactions on Biomedical Engineering 54(11):1940–1950
8. Nakamura Y, Yamane K, Fujita Y, Suzuki I (2005) Somatosensory computation for man-machine interface from motion-capture data and musculoskeletal human model. IEEE Transactions on Robotics 21(1):58–66
9. Zajac FE (1989) Muscle and tendon: Properties, models, scaling and application to biomechanics and motor control. CRC Critical Reviews in Biomedical Engineering 17:359–411
10. Wieber PB (2000) Modélisation et commande d'un robot marcheur anthropomorphe. PhD Thesis, Ecole des mines de Paris
11. Bottecchia S (2006) Estimation de posture : vers la prise en compte de la 3° dimension sur le modèle biomécanique. Rapport de stage de Master II Recherche, Lirmm, Montpellier
12. Hayashibe M, Poignet P, Guiraud D, Makssoud H (2008) Nonlinear identification of skeletal muscle dynamics with sigma-point Kalman filter for model-based FES. In: Proceedings of the IEEE International Conference on Robotics and Automation, pp. 2049–2054

Hierarchical Markov Random Fields Applied to Model Soft Tissue Deformations on Graphics Hardware

Christof Seiler[1], Philippe Büchler[2], Lutz-Peter Nolte[3], Mauricio Reyes[4], and Rasmus Paulsen[5]

Abstract Many methodologies dealing with prediction or simulation of soft tissue deformations on medical image data require preprocessing of the data in order to produce a different shape representation that complies with standard methodologies, such as mass–spring networks, finite element methods (FEM). On the other hand, methodologies working directly on the image space normally do not take into account mechanical behavior of tissues and tend to lack physics foundations driving soft tissue deformations. This chapter presents a method to simulate soft tissue deformations based on coupled concepts from image analysis and mechanics theory. The proposed methodology is based on a robust stochastic approach that takes into account material properties retrieved directly from the image, concepts from continuum mechanics and FEM. The optimization framework is solved within a hierarchical Markov random field (HMRF) which is implemented on the graphics processor unit (GPU).

Introduction

One of the current problems with non-rigid registration techniques is their lack of physics foundations concerning mechanical properties and energies driving the deformations. Conversely, classical methods to compute soft tissue deformations

[1] University of Bern, ARTORG Center, 3014 Bern, Switzerland
christof.seiler@artorg.unibe.ch

[2] University of Bern, ARTORG Center, 3014 Bern, Switzerland
christof.seiler@artorg.unibe.ch

[3] University of Bern, ARTORG Center, 3014 Bern, Switzerland
christof.seiler@artorg.unibe.ch

[4] University of Bern, ARTORG Center, 3014 Bern, Switzerland
christof.seiler@artorg.unibe.ch

[5] Technical University of Denmark, DTU Informatics, 2800 Lyngby, Denmark
rrp@imm.dtu.dk

N. Magnenat-Thalmann et al. (eds.), *Recent Advances in the 3D Physiological Human*,
DOI 10.1007/978-1-84882-565-9_9,

such as finite element methods (FEM), mass–spring networks. suffer from the need of transforming the data into a volumetric mesh representation of the object. As a consequence, image segmentation and volumetric mesh generation are two inevitable steps to be performed. Whereas image segmentation can be performed semi- or fully automatically with some acceptable accuracy, automatic volumetric meshing algorithms are prone to errors and dependent on object topology, desired boundary conditions [1], etc. Therefore, semi-manual techniques are often used, which is tedious and time consuming. Furthermore, jumping from the image space (i.e., voxels) to the shape space (i.e., mesh) involves approximations due to the discrete and different nature of both data representations [15]. In addition, voxel intensities have valuable information about material properties of the structure, which is required to be translated to the shape space (linear and non-linear interpolations, barycentric-based techniques, etc.) at the price of introducing further errors due to these approximations. The aforementioned limitations of these two approaches (mesh- and image-based techniques) are known by the scientific community and efforts have been made in this direction. In [2, 17] automatic FEM meshing techniques are proposed. The method generates a one-to-one representation of voxels into hexahedra elements and a posterior smoothing approach is applied to deal with jagged edges of the geometry. However, the high number of generated elements yield long computations making this technique only applicable to micro-CT images. In [11] an approach is presented for deformable registration of brain tumor images to a normal brain atlas, here the dissimilarity of the images hinders the usability of readily available deformable image registration techniques. Through statistical modeling of sought tumor-induced deformations the method combines a biomechanical model of tumor mass effect and a deformable image registration technique. By doing this, the authors reported significant reduction in the registration error. Nonetheless, the reliability of the method depends on the statistical training performed on a set of FEM simulations using different parameters of the tumor mass effect, such as tumor sizes, locations, external pressures. Inspired by the work presented in [13] here we present a method to simulate soft tissue deformations in the image space that is based on a maximum a posteriori (MAP) model of the deformations, which considers novel developed energy terms to account for tissue material properties, boundary conditions, and related confidence maps. Stochastic optimization is performed under a Markov random field (MRF) approach, which is further extended into a hierarchical MRF (HMRF) approach. The use of a HMRF approach enables us to use fast local optimizers, as opposed to global optimizers which are computationally very expensive. The hierarchical approach is also robust with respect to local minima [14]. Finally, MRF and HMRF are well suited for parallel implementation [9, 10], which has brought more attention with the advent of GPU-based application programming interfaces (API).

Test Data

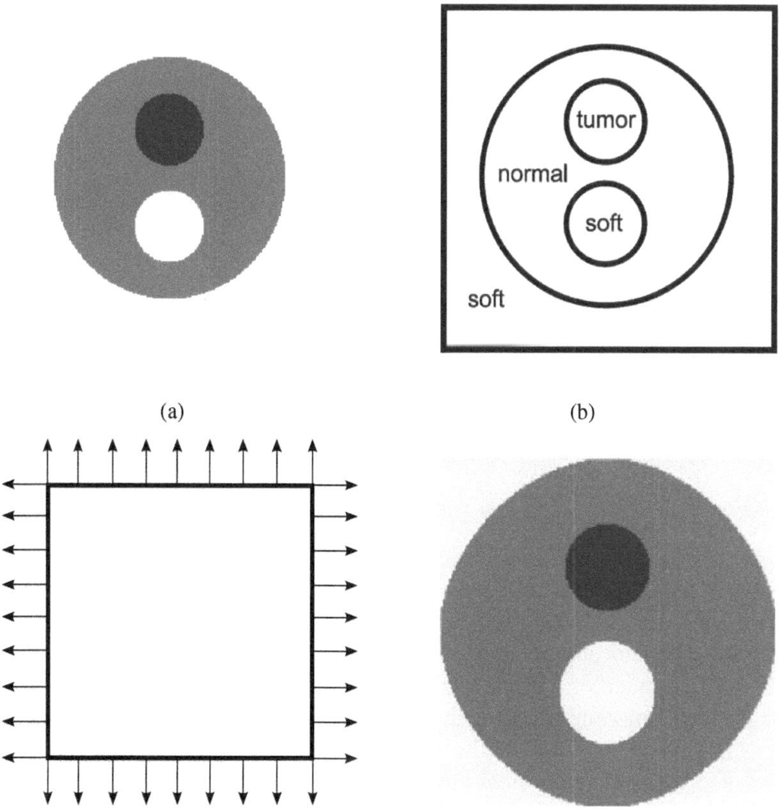

(a) (b)

(c) (d)

Fig. 1. (**a**) Input image, (**b**) segmentation of the input image, (**c**) representation of boundary condition image with outward expansion, and (**d**) deformed input image.

A set of synthetic data are used to validate the method. These data are generated so it resembles an MRI image of a brain, as seen in Fig. 1a. To simplify notations three tissue types are considered. Namely, normal matter (gray color), tumoral tissue (black color), and soft tissue (white color), depicted in Fig. 1b. Second, a boundary condition image is constructed as seen in Fig. 1c, where the displacement vectors are shown. In the following, the boundary condition image is described by $\mathbf{y}$. In this example, the boundary condition image describes an outward expansion. In real applications, the boundary condition image can, for example, be computed using a non-rigid registration algorithm [3, 16]. To be able to control the amount of confidence in the boundary condition image a confidence image is also provided. In the following, the confidence image is described by **c**. In this implementation, it is a binary vector-valued image where a value of 0 means no confidence in the boundary condition image at that pixel and a value of 1 means complete confidence. In the presented example the confidence image coincides with Fig. 1c except that the vectors are normal vectors. The use of confidence fields is described in [12]. Finally, an image **m**, describing the local material properties, is used. In our example, it corresponds to the input image where each pixel is assigned the Young's modulus stiffness based on the underlying tissue. A value of 40 MPa is used for tumors, 20 MPa for the normal tissue, and 10 MPa for soft tissue. In a real application, the mechanical properties can be assigned using a segmentation and classification algorithm. In conclusion, the following images are used as input for the proposed method: segmented input image, boundary condition image **y**, confidence image **c**, and mechanical properties image **m**.

Methods

The aim of our method is to compute a displacement field that conforms both to a given boundary condition image and a set of mechanical properties of the underlying tissue. In the following it is demonstrated how this is solved using Markov random field (MRF) regularization. Using MRF enables us to describe the underlying mechanical properties of the tissue using a prior energy term and to set the boundary conditions using an observation energy term. Because no initial guess for the displacement vector field is available the field is set to null vectors at all sites.

The displacement vector field is described using a multivariate random variable **D**. A realization of the vector field is described by **d**. In addition, each vector in the displacement field is also described by a multivariate random variable $\mathbf{d}_s$, where s denotes the spatial location of the vector. This will be explained in more detail in the next sections.

Maximum A Posteriori and Markov Random Fields

The final objective is to find a vector displacement field $\hat{\mathbf{d}}$ that maximizes the posterior probability P,

$$\hat{\mathbf{d}} = \arg\max_{\mathbf{d}} P(\mathbf{d} \mid \mathbf{y}), \tag{1}$$

given a boundary condition image $\mathbf{y}$. MRFs are defined on a finite index set S with elements s called sites. Random variables are associated to each element in S. To be able to compute the posterior probability the theorem of Hammersley–Clifford is used, which establishes the equivalence between MRF and Gibbs distribution, a proof can be found in [4, 18]. With the help of the Gibbs distributions

$$P(\mathbf{d}) = \frac{1}{Z}\exp(-U(\mathbf{d})/T), \tag{2}$$

where Z, U, and T are the normalizing constant, the energy function, and the temperature, respectively; the probability can be defined in terms of energy functions. Energy functions are the summation of potential functions:

$$U(\mathbf{d}) = \sum_{C \in \mathsf{C}} V_C, \tag{3}$$

where V_C are defined on local neighborhood systems N. A local neighborhood system is a subset of S, usually only a few sites are involved. C is called a clique if any two different elements of C are neighbors, so C is a subset of N. The set of cliques will be denoted by C. The Markov property of MRF is expressed as [5]

$$P(\mathbf{d_s} \mid \mathbf{d_t}, t \neq s) = P(\mathbf{d_s} \mid \mathbf{d_t}, t \in N_s) \forall s \in S, \tag{4}$$

where N_s is a neighborhood system at site s, $\mathbf{d_s}$ and $\mathbf{d_t}$ are vectors at site s and t, respectively. It means that the probability distribution of each site only depends on the state of neighboring sites. In Bayesian terms the a posteriori probability can be written as

$$P(\mathbf{d} \mid \mathbf{y}) = \frac{P(\mathbf{y} \mid \mathbf{d})P(\mathbf{d})}{P(\mathbf{y})}. \tag{5}$$

In image processing applications $P(\mathbf{d})$ stands for the a priori energy term, which represents prior knowledge about the configuration of $\mathbf{d}$. $P(\mathbf{y}|\mathbf{d})$ represents the observation function, which describes the relation between $\mathbf{y}$ and $\mathbf{d}$. $P(\mathbf{y})$ is the density of $\mathbf{y}$, which is a constant when $\mathbf{y}$ is given. The combination of (2), (3), and (5) yields

$$P(\mathbf{d} \mid \mathbf{y}) \propto \exp(-U_{\text{total}}/T), \tag{6}$$

where

$$U_{\text{total}} = U_{\text{prior}} + U_{\text{observation}} \tag{7}$$

is the sum of observation and prior energy. To maximize (6) one can minimize (7).

Prior Energy

In the current implementation, the prior energy is based on the mechanical properties of the underlying tissue. The prior energy should therefore be formulated so it is at its minimum when the deformation of the tissue conforms to the expected mechanical properties. Mechanical properties of the tissue are modeled using a finite difference approach, where the local tissue characteristics are based on Young's modulus. The used energy function is then

$$U_{\text{prior}}(\mathbf{d},\mathbf{m}) = \sum_{C_1 \in C} V_{C_1} + \sum_{C_2 \in C} V_{C_2} + \sum_{C_3 \in C} V_{C_3} + \sum_{C_4 \in C} V_{C_4}, \tag{8}$$

where C_1, C_2, C_3, and C_4 are four different kinds of cliques, each involving three pixels, which can be visualized as in Fig. 2.

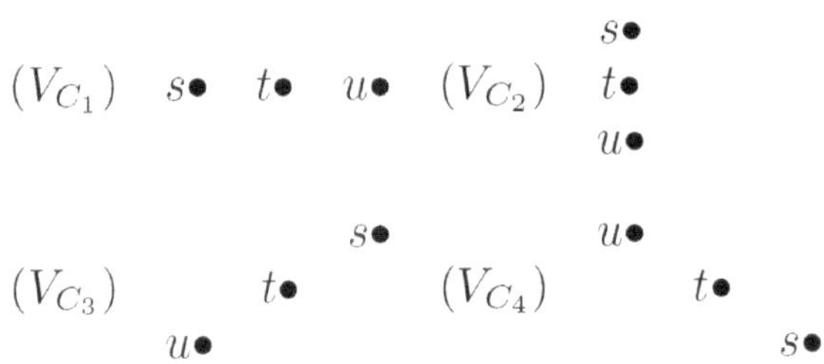

Fig. 2. Four cliques for the prior energy.

The four types of cliques are assigned with potential functions to describe the biomechanical process of deformation:

$$V_{C_1} = \left(\frac{m_t}{m_u}(d_{x_t} - d_{x_s}) - (d_{x_u} - d_{x_t}) \right)^2 + \left(\frac{m_t}{m_s}(d_{x_t} - d_{x_u}) - (d_{x_s} - d_{x_t}) \right)^2 \tag{9}$$

$$V_{C_2} = \left(\frac{m_t}{m_u}(d_{y_t} - d_{y_s}) - (d_{y_u} - d_{y_t}) \right)^2 + \left(\frac{m_t}{m_s}(d_{y_t} - d_{y_u}) - (d_{y_s} - d_{y_t}) \right)^2 \tag{10}$$

$$V_{C_{3,4}} = |\underbrace{\frac{m_t}{m_u}(\mathbf{d_t} - \mathbf{d_s})}_{(a)} - \underbrace{(\mathbf{d_u} - \mathbf{d_t})}_{(b)}|^2 + |\underbrace{\frac{m_t}{m_s}(\mathbf{d_t} - \mathbf{d_u})}_{(c)} - \underbrace{(\mathbf{d_s} - \mathbf{d_t})}_{(d)}|^2, \quad (11)$$

where $\mathbf{d_i} = [\mathbf{d_{xi}}\ \mathbf{d_{yi}}]^T$ and m_i are the displacement vector at site i and the biomechanical property image sampled at i, respectively. Indicated by letters (a) to (d) there are four parts that define the prior knowledge, analog for (9) and (10). Parts (a) and (b) describe the change of Young's modulus in one direction and (c) and (d) the change in the opposite direction. The goal of the prior energy is to reach

$$\frac{m_t}{m_u}(\mathbf{d_t} - \mathbf{d_s}) = (\mathbf{d_u} - \mathbf{d_t}) \quad (12)$$

at every site. The relative change at site t is equal to the relative change at u by a ratio of m_t/m_u. The differences $(\mathbf{d_t}-\mathbf{d_s})$ and $(\mathbf{d_u}-\mathbf{d_t})$ describe the elasticity in terms of relative displacements at site t and u, respectively. To assign an energy value to the relative difference this translates to

$$\left|\frac{m_t}{m_u}(\mathbf{d_t} - \mathbf{d_s}) - (\mathbf{d_u} - \mathbf{d_t})\right|^2. \quad (13)$$

This means that the minimum energy at site t is given by 0. The bigger the difference the higher the energy. Also by taking the square of the energy value, values >1 contribute more than values <1. The energy function also detects non-monotonic fields. A monotonic field is defined as

$$|\,\mathbf{d_s}\,| \le |\,\mathbf{d_t}\,| \le |\,\mathbf{d_u}\,| \quad (14)$$

or

$$|\,\mathbf{d_s}\,| \ge |\,\mathbf{d_t}\,| \ge |\,\mathbf{d_u}\,|. \quad (15)$$

Example 1. [Zero energy at site t] In this example a situation is presented where two neighboring pixel have different Young's modulus values. The minimum energy is reached when (12) is fulfilled. A possible configuration for three displacement vectors $\mathbf{d_s}$, $\mathbf{d_t}$, and $\mathbf{d_u}$ is given.

$$\mathbf{d_s} = \begin{pmatrix} 1 \\ 1 \end{pmatrix} \text{ mm}, \ \mathbf{d_t} = \begin{pmatrix} 2 \\ 2 \end{pmatrix} \text{ mm}, \ \mathbf{d_u} = \begin{pmatrix} 4 \\ 4 \end{pmatrix} \text{ mm}, \ m_t = 20 \text{ MPa}, \ m_u = 10 \text{ MPa}$$

$$\frac{20}{10}\left(\begin{pmatrix} 2 \\ 2 \end{pmatrix} - \begin{pmatrix} 1 \\ 1 \end{pmatrix}\right) = \left(\begin{pmatrix} 4 \\ 4 \end{pmatrix} - \begin{pmatrix} 2 \\ 2 \end{pmatrix}\right)$$

Example 2. [Non-monotonic field at site t] In the second example both pixels have the same Young's modulus. Inserting the example values into (13) gives the energy value.

$$\mathbf{d_s} = \begin{pmatrix} 1 \\ 1 \end{pmatrix} \text{ mm}, \ \mathbf{d_t} = \begin{pmatrix} 2 \\ 2 \end{pmatrix} \text{ mm}, \ \mathbf{d_u} = \begin{pmatrix} 1 \\ 1 \end{pmatrix} \text{ mm}, \ m_t = 10 \text{ MPa}, \ m_u = 10 \text{ MPa}$$

$$\left|\frac{10}{10}\left(\begin{pmatrix} 2 \\ 2 \end{pmatrix} - \begin{pmatrix} 1 \\ 1 \end{pmatrix}\right) - \left(\begin{pmatrix} 1 \\ 1 \end{pmatrix} - \begin{pmatrix} 2 \\ 2 \end{pmatrix}\right)\right|^2 = 2.8$$

Even though the difference between sites and the biomechanical properties are the same, a non-zero value is calculated. Non-monotonic fields cause overlapping of sites, which is not realistic.

Observation Energy

The current application uses the observation energy to set the boundary conditions:

$$U_{\text{observation}}(\mathbf{y}, \mathbf{c}, \mathbf{d}) = \sum_{C_5 \in \mathrm{C}} V_{C_5} + \sum_{C_6 \in \mathrm{C}} V_{C_6}, \tag{16}$$

where C_5 and C_6 are cliques at single sites s and

$$V_{C_5} = p_{x_s} \mid y_{x_s} - d_{x_s} \mid \tag{17}$$

$$V_{C_6} = p_{y_s} \mid y_{y_s} - d_{y_s} \mid, \tag{18}$$

where **p** is the penalty vector field image. The penalty image is proportional to the confidence image

$$p_{x_s} = Kc_{x_s}, \quad p_{y_s} = Kc_{y_s}, \tag{19}$$

where $\mathbf{c}_s = [c_{xs}\ c_{ys}]^T$, $\mathbf{p}_s = [p_{xs}\ p_{ys}]^T$, and K is the penalty constant. By issuing a penalty, a low energy value is sensitive to the difference between the boundary vector and the displacement vector. Hence, the probability is high when the displacement vector at a site coincides with the boundary condition vector whenever boundary conditions are available, indicated by the confidence vector.

Hierarchical Markov Random Fields

The idea is to propagate the geometrical information through hierarchical levels. Low-resolution scale levels express global geometrical features whereas high-resolution scale levels are used for fine-tuning. Geometrical information is needed to converge to a global solution. By using a hierarchical approach, local energy optimizers can be used to find global solutions. Local optimizers are stable and predictable whereas global optimizers do not guarantee convergence and the computation time is enormous, this is further discussed in the next section.

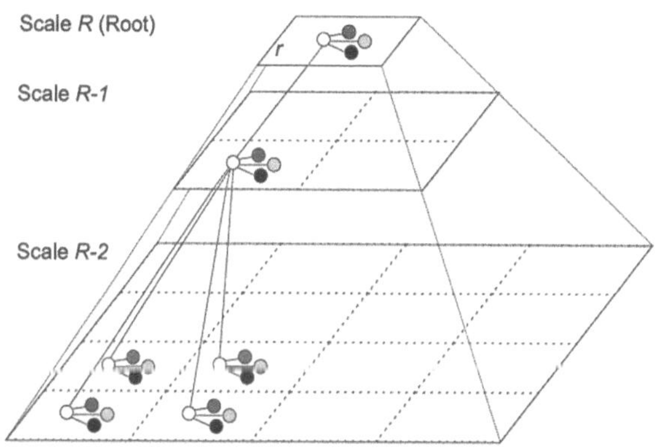

Fig. 3. Dependency graph corresponding to a quadtree structure, with *white*, *light gray*, *dark gray*, and *black circles* representing displacement vectors, boundary condition vectors, confidence vectors, and mechanical property vectors, respectively.

In [14] HMRF for segmentation is presented; it follows the adaptation to the presented method. Let $G = (S,L)$ be a graph composed of a set S of nodes and a set L of edges. A tree is a connected graph that has no cycle. Each node has a unique parent node, except the root node r. A quadtree, as illustrated in Fig. 3, is a special case of a tree. Each node in a quadtree has four child nodes except the leave nodes, which are terminal nodes. Set S can be partitioned into subsets we call *scale levels*. The subset, $S = S^0 \cup S^1 \cup \cdots \cup S^R$, are distinguished according to the path length from each node to the root. S^0 is the subset with the most elements and $S^R = \{r\}$ the one with the least elements. A subset contains 4^{R-n} sites, where R represents the scale and n the distance from the root node r.

Different interpolation methods are used for projecting the images from one scale level to the next. **d** is interpolated linearly, because linear elastic behavior was used to describe the material properties of the image. On the other hand, for the interpolation of **m**, **y**, and **c**, nearest neighbor interpolation is used to preserve information at interfaces between different tissue types and prevent over-smoothing. In Fig. 3 the propagation between scale levels is shown, starting on top with the root element r. To each white circle $\mathbf{d}_s \in \mathbf{d}$, three circles $\mathbf{m}_s \in \mathbf{m}$, $\mathbf{y}_s \in \mathbf{y}$, and $\mathbf{c}_s \in \mathbf{c}$ are attached, representing interpolated pixel values at corresponding locations s.

Energy Minimization

In practice, the crucial point of optimization is a good initial guess. With initial guesses far off the global minimum, local energy optimizers like iterated conditional modes (ICM) [18] will most likely stay at a local minimum and global energy optimizers like Markov chain Monte Carlo (MCMC) with simulated annealing (SA) [18] methods will take a long time; if the cooling schedule is not adjusted to the problem, it might even stop at a local minimum as well. In the current approach, the overall geometrical structure of tissues and their corresponding mechanical properties are crucial in obtaining a good initial guess. A top-down hierarchical approach is used to supplement the local nature of the used energy terms. Each level of the hierarchy is used as an initial guess for the next level. Iterated conditional modes is used as local optimizer within each hierarchy level because a good initial guess from the previous level can be assumed.

Hierarchical Markov Random Fields on GPU

Even though ICM is used as optimization method, computation of the entire quadtree is expensive. As will be explained in the following sections around 70 iterations per scale level are needed to reach convergence. By implementing the method on graphics hardware the computation time can be reduced drastically. Due to the fact that MRF is only looking at neighborhood information it is fairly easy to implement it in a highly parallelized way.

Implementation

In [9, 10] the general way of MRF on graphics hardware is presented. For the implementation a new extension for OpenGL, EXT_framebuffer_object [6, 7], was used. Code execution on the GPU is done through shader programs. The shader program is executed on each pixel simultaneously. As input arguments the shader program has access to read-only memory; for output values the shader program has access to write-only memory. The EXT_framebuffer_object enables alternating between read and write buffers directly without copying the buffers back and forth after one iteration has been completed. Two textures are created as read/write data structures on the GPU. One framebuffer object is created and the two textures are attached. Using the glDrawBuffer extension, ping-pong rendering is implemented. Each scale level is computed until the finest scale level is reached.

The current implementation has two restrictions: First, the number of iterations needs to be predefined before the simulation is started. Second, one can only read or write to memory and not both, this means that we cannot update the image during the iteration process.

The pseudocode for the implementation is listed in Alg. 3.7. U_{total} is calculated according to (7) and Γ is a set of values describing the next estimate, e.g., Γ={(0.1,-01),(0.001,0.001),…}. By varying Γ one can set the step size for the optimization method. Lines 3 to 6 are the projection from one scale level to the next. Lines 7 to 10 are the ICM minimization. The loop will continue until the full resolution image is reached.

Computation on the quadtree (with notations from Fig. 3) and optimization with ICM:

Data: boundary condition image **y**, confidence image **c**, mechanical property image **m**

Result: displacement vector field $\hat{\mathbf{d}}$

initialize **d** to null vector field
for n−R to 0 **do**

- projection of **d** from scale n+1 to n by linear interpolation
- projection of **y** from scale 0 to n by nearest neighbor interpolation
- projection of **c** from scale 0 to n by nearest neighbor interpolation
- projection of **m** from scale 0 to n by nearest neighbor interpolation
- **repeat**
- **foreach** site $s \in S$ **do**
- $\mathbf{d}_s \leftarrow \arg\min_{e \in \Gamma} (U_{total}(\mathbf{y}_s, \mathbf{c}_s, \mathbf{d}_s + e, \mathbf{m}_s))$
- **until** U_{total} stabilizes

Experiments and Results

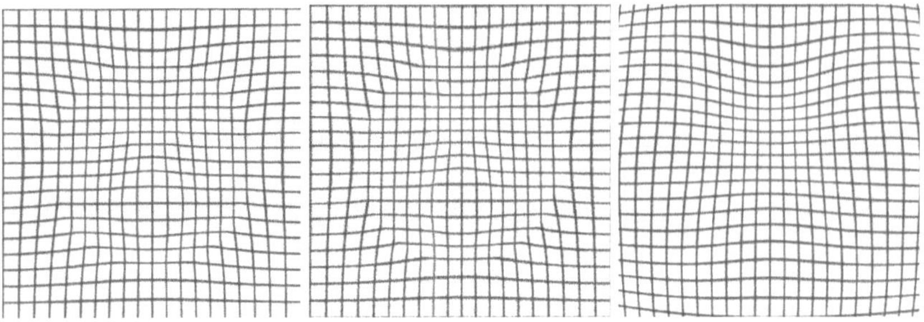

Fig. 4. Deformation grids: (a) HMRF method, (b) ABAQUS, and (c) elastic registration method.

In general non-rigid deformation algorithms do not include mechanical property information which results in non-realistic deformation. Therefore, a comparison with an existing elastic deformation registration method [3] was made. As will be explained in Section Ground Truth, a FEM model was taken as the ground truth deformation. The elastic registration method was used to register the deformation image received from ABAQUS[6] with the input image. The resulting displacement vector fields are shown as deformation grids in Fig. 4a–c. The HMRF model and ABAQUS show similar deformations whereas the elastic registration method does not consider mechanical properties, which can be seen by looking at tissue interfaces. The elastic registration image looks smooth because the change of biomechanical properties at the interfaces is not taken into account.

Ground Truth

To validate the results of the simulation the same mechanical deformation in a FEM model in ABAQUS was implemented and used as ground truth image. A comparison of data using a root mean square deviation (RMSD) was realized,

$$\mathrm{RMSD}(x) = \sqrt{\frac{1}{|S|}\sum_{s\in S} |\mathbf{a}_s - \hat{\mathbf{d}}_s|^2}, \tag{20}$$

where $\mathbf{a}_s$ and $\hat{\mathbf{d}}_s$ are the displacement vector obtained from ABAQUS and our model, respectively.

[6]ABAQUS is a software tool to create FEM models

Quality

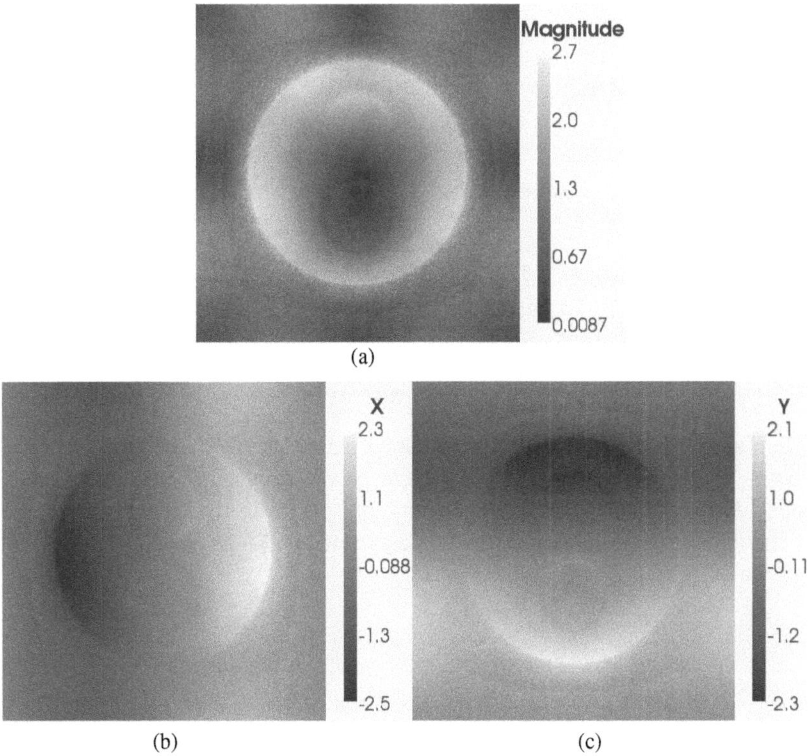

Fig. 5. ABAQUS and HMRF displacement vector field images differences: a) magnitude of the vector b) *x* component of the vector and c) *y* component of the vector.

Table 1. Experimental data obtained to evaluate quality compared to FEM, 4255 elements of type CPS3 (3-node linear) and CPS4R (4-node bilinear, reduced integration with hourglass control) are used for the FEM calculations.

Method	Mean energy value	RMSD error FEM - HMRF	Iterations	Computation time
HMRF	0.19404	1.44 pixel	700	1.63 s
FEM	-	-	-	1.30 s

The comparison with FEM results are shown in Table 1. In Fig. 5a the difference image with ABAQUS results is shown. It can be seen that the regions that differ the most are around the outer corner of the synthetic brain. Further it is interesting to look at the two figures (Fig. 5b, c) in which the difference per component is depicted. In both images the results show that most of the differences occur at the outer border of the synthetic brain, which confirms the interpretation of Fig. 5a.

Computation Time

Table 2. Comparison of computation times between GPU and CPU implementation with 70 iteration steps.

Image size	GPU time(s)	CPU time(s)	Speed-up
300 × 300	1.63	91.39	56.07
600 × 600	6.42	363.89	56.68
1200 × 1200	26.25	1445.94	55.08

In Table 2 a listing on the performance gain with GPU compared to CPU is shown. Three different image sizes have been examined, a speed-up factor of around 55 was found. The number of iterations per scale level is set to 70.

The experiments were performed on a system with ATI, Radeon X1300 Series, and Intel, Core™ 2 Duo E6300 1.86 GHz.

Discussion and Future Work

A method to simulate soft tissue deformations in the image space using physical-based concepts and image processing techniques was presented. The method combines mechanical concepts into a Bayesian optimization framework which has been defined and solved under a HMRF approach and implemented for the GPU. The main advantage of the proposed methodology over previous ones is its mesh-free characteristic, which is normally needed to perform accurate mechanical simulations of tissues. The use of HMRF proves to be an appealing technique to solve the proposed stochastic problem. The reasoning is twofold: On one hand it was found that local minima are avoided by using a hierarchical approach, which in turn allows for a speed-up since it is now possible to use local optimizers rather than slow global optimizers (e.g., Monte Carlo optimization). Second, the nature of the Markovian approach resulted in a straightforward implementation in the GPU, which has been remarked by others [9, 10]. The experiments showed that by using HRMF at the GPU level, stable results can be attainable in a very reasonable computation time, simplifying the soft-tissue-deformation simulation pipeline, making this method very appealing for medical applications.

In future experiments the focus will be on more complex geometry and non-synthetic image data. In these regards ongoing work includes its extension to 3D models. An implementation with OpenGL and Cg shader programming was realized. In the future a transfer to the Brook [8], project from Stanford University, or CUDA, NVIDIA frameworks will be considered. A strong influence of the number of hierarchical levels on the number of iterations needed for the model to converge is assumed, this topic will also be further investigated in the future.

References

1. Bachtar F, Chen X, Hisada T (2006) Finite element contact analysis of the hip joint. Medical and Biological Engineering and Computing 44(8):643–651
2. Boyd SK, Müller R (2006) Smooth surface meshing for automated finite element model generation from 3D image data. Journal of Biomechanics 39(7):1287–1295
3. Carreras IA, Sorzano C, Marabini R, Carazo J, De Solorzano CO, Kybic J (2006) Consistent and Elastic Registration of Histological Sections Using Vector-Spline Regularization. Computer Vision Approaches to Medical Image Analysis, pp. 85–95
4. Geman D (1990) Random Fields and Inverse Problems in Imaging. In: Saint-Flour Lectures 1988, Lecture Notes in Mathematics, pp. 113–193
5. Geman S, Geman D (1984) Stochastic relaxation, Gibbs distributions and the Bayesian restoration of images. IEEE Transactions on Pattern Analysis and Machine Intelligence 6(6): 721–741
6. Göddeke D (2005) GPGPU Basic Math Tutorial. Technical report, FB Mathematik, Universität Dortmund
7. Green S (2005) The OpenGL Framebuffer Object Extension. NVIDIA Corporation, GDC
8. Gummaraju J, Rosenblum M (2005) Stream Programming on General-Purpose Processors. In: Proceedings of the 38th annual IEEE/ACM International Symposium on Microarchitecture, Barcelona, Spain
9. Jodoin PM, Mignotte M (2006) Markovian segmentation and parameter estimation on graphics hardware. Journal of Electronic Imaging 15(3):033005-1–033005-15
10. Jodoin PM, St Amour JF, Mignotte M (2005) Unsupervised Markovian segmentation on graphics hardware. Pattern Recognition and Image Analysis 3687:444–454
11. Mohamed A, Zacharaki EI, Shen D, Davatzikos C (2006) Deformable registration of brain tumor images via a statistical model of tumor-induced deformation. Medical Image Analysis 10(5):752–763
12. Murino V, Castellani U, Fusiello A (2001) Disparity map restoration by integration of confi dence in Markov random fields models. In: International Conference on Image Processing
13. Paulsen RR (2004) Statistical Shape Analysis of the Human Ear Canal with Application to In-the-Ear Hearing Aid Design. PhD thesis, Technical University of Denmark
14. Provost JN, Collet C, Rostaing P, Pérez P, Bouthemy P (2004) Hierarchical Markovian segmentation of multispectral images for the reconstruction of water depth maps. Computer Vision and Image Understanding 93(2):155–174
15. de Putter S, van de Vosse FN, Gerritsen, FA, Laffargue F, Breeuwer M (2006) Computational mesh generation for vascular structures with deformable surfaces. International Journal of Computer Assisted Radiology and Surgery 1(1):39–49
16. Rueckert D, Sonoda LI, Hayes C, Hill DLG, Leach MO, Hawkes DJ (1999) Nonrigid registration using free-form deformations: Application to breast MR images. IEEE Transactions on Medical Imaging 18(8):712–721

17. Sigal IA, Hardisty MR, Whyne CM (2008) Mesh-morphing algorithms for specimen-specific finite element modeling. Journal of Biomechanics, 1381–1389.
18. Winkler G (2006) Image Analysis, Random Fields and Markov Chain Monte Carlo Methods. Springer, Second edition

A Physics-Based Modeling and Real-Time Simulation of Biomechanical Diffusion Process Through Optical Imaged Alveolar Tissues on Graphical Processing Units

Ilhan Kaya[1], Anand P. Santhanam[2], Kye-sung Lee[3], Panomsak Meemon[4], Nicolene Papp[5], and Jannick P. Rolland[6]

Abstract Tissue engineering has broad applications from creating the much-needed engineered tissue and organ structures for regenerative medicine to providing in vitro testbeds for drug testing. In the latter, application domain, creating alveolar lung tissue, and simulating the diffusion process of oxygen and other possible agents from the air into the blood stream as well as modeling the removal of carbon dioxide and other possible entities from the blood stream are of critical importance to simulating lung functions in various environments. In this chapter, we propose a physics-based model to simulate the alveolar gas exchange and the alveolar diffusion process. Tissue engineers, for the first time, may utilize these simulation results to better understand the underlying gas exchange process and properly adjust the tissue growing cycles. In this work, alveolar tissues are imaged by means of an optical coherence microscopy (OCM) system developed in our laboratory. As a consequence, 3D alveoli tissue data with its inherent complex boundary is taken as input to the simulation system, which is based on computational fluid mechanics in simulating the alveolar gas exchange. The visualization and the simulation of diffusion of the air into the blood through the alveoli tissue is performed using a state-of-art graphics processing unit (GPU). Results show the real-time simulation of the gas exchange through the 2D alveoli tissue.

[1] University of Central Florida, CREOL, Orlando FL, USA
ikaya@creol.ucf.edu

[2] University of Central Florida, CREOL, Orlando FL, USA
M.D. Anderson Cancer Center Orlando, Orlando FL, USA
asanthan@creol.ucf.edu

[3] University of Rochester, Rochester NY, USA
lee@optics.rochester.edu

[4] University of Central Florida, CREOL, Orlando FL, USA
pmeemon@creol.ucf.edu

[5] University of Central Florida, CREOL, Orlando FL, USA
nwpapp@knights.mail.ucf.edu

[6] University of Rochester, Rochester NY, USA
rolland@optics.rochester.edu

N. Magnenat-Thalmann et al. (eds.), *Recent Advances in the 3D Physiological Human*,
DOI 10.1007/978-1-84882-565-9_10,

Introduction

Tissue engineering is a collaborative effort that transcends different fields of science and technology: biology, mechanics, mathematics, computer science, and engineering. It is of utmost important that the engineered tissues replicate the normal functionality of the tissue being replaced. Thus the modeling and simulation of the tissue's biomechanics and its real-time visualization will enable tissue engineering experts to understand the engineered tissue's functionality and to estimate their functionality under different physiological conditions.

Our focus is on the biomechanics of the lung epithelial tissue and its role in the alveolar gas exchange process, which plays a key role in oxygenating the blood and releasing the carbon dioxide from the blood. Recent technological advances in medical imaging have led to the availability of 4D macroscopic lung images and the estimation of macroscopic lung tissue properties such as the Young's modulus [1, 2]. However, the complexity of imaging the microstructures inside the lungs has limited the understanding of the alveolar gas exchange process in a single breath. Thus modeling and simulating the alveolar tissue's role in the gas exchange process will shed light on tissue engineering-related issues.

Commercial software such as CFD-ACE and CFD-GEOM enable a generalized tool for performing computational fluid dynamics for both smooth and complex boundaries [3]. These tools are numerically precise in the computations but are limited by their non-real-time performance, which hampers a real-time visualization of the gas exchange process. A seminal work on simulating stable fluids within smooth boundaries was proposed by Stam et al. [4, 5], who employed a semi-Lagrangian method invented by Courant et al. [6]. This work introduced the stable real-time CFD simulations to the graphics and animation community. More recent work done by peers in this community has shown that accurate fluid dynamics simulations for stable fluids and fluid–solid couplings within smooth boundaries are possible [7, 8]. The simulation of stable fluids within accurate complex boundaries has not been investigated in the graphics community because of its computational complexity. The complexity in simulating the computational fluid dynamics is alleviated by the usage of the state-of-art graphics processing units (GPU) that provides a computational capability of approximately 1 Tera FLOPS. Such GPUs have been employed for simulating 3D fluid flow dynamics for simple boundaries [9] and 3D cloud dynamics [10].

In this chapter, we are extending the framework discussed in [4] to simulate the alveolar gas exchange through the complex and permeable tissue boundaries on GPUs. The diffusion of oxygen and carbon dioxide in the tissue requires a crossing of the boundary in the tissue–air and tissue–blood boundaries. A 3D tissue membrane is obtained from a cultured epithelial tissue, which is imaged in our laboratory through an optical imaging technique known as optical coherence microscopy (OCM) [11]. The complex boundary is discretized to a fine scale, thereby facilitating an accurate representation of the alveolar tissue surface for the simulation system. Moreover, the real-time simulation of the alveolar gas exchange and diffusion through the tissue provides an effective visualization of the underlying process for the tissue engineering experts. The key contribution of this chapter is the inclusion of complex boundaries with porous membrane property for simulating the gas exchange in addition to the permeability and the diffusivity of air through the permeable membrane into the blood on GPUs.

This chapter is organized as follows: In the next section we briefly describe the optical image acquisition process. Then, we present the theory and the mathematical model that we used to simulate the alveolar gas exchange and diffusion processes through a complex boundary. Subsequent to the description of the mathematical model, the implementation details and issues regarding the numerical methods and complex boundaries are explained. We conclude the chapter with 2D results showing the diffusion process in the permeable alveolar tissue constructs.

Optical Imaging of Alveolar Tissues by Optical Coherence Microscopy (OCM)

Optical coherence microscopy (OCM) is a noninvasive high-resolution and high-sensitivity depth-resolved imaging technique [11], which is developed in our laboratory for high invariant resolution throughout the sample [12]. By measuring the backscattering and back-reflected light, OCM systems provide cross-sectional images of the internal microstructures presented in the alveolar tissues. The OCM system used in our laboratory to image the alveoli tissues is depicted in Fig. 1.

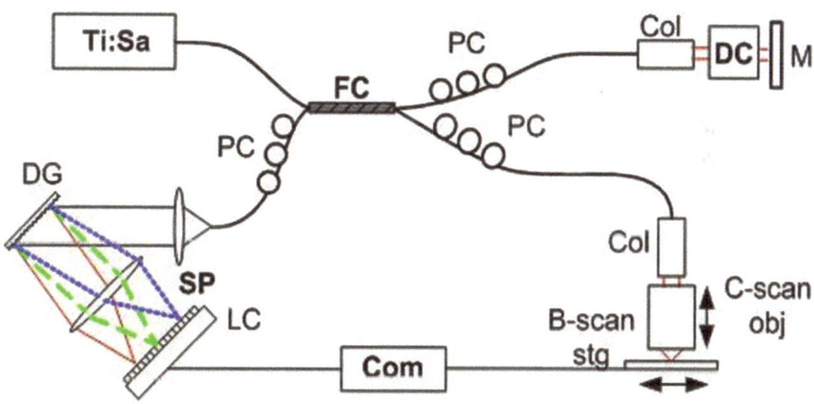

Fig. 1. A schematic of the custom-built OCM system.

In Fig. 1, FC represents a fiber coupler, PC are polarization controllers, Col is collimating optics, DC is the dispersion compensation mechanism, DG is a diffraction grating, LC is a line CCD, SP is a spectrometer, and M is a mirror. An OCM system is similar in essence to an ultrasound imaging system in the sense that the time that the light reflects off the tissue is measured. Since light propagates too fast to be measured directly, an interferometer is used to measure relative time.

By collecting 200 axial scans and 200 x–z images with OCM we have gathered 200 μm × 200 μm surface and 25 μm depth information of the alveolar tissue in 3D. The 3D data are then visualized through a custom-built volume renderer employing a 3D texture slicing technique on GPU [13]. A 3D visual volumetric representation of the alveolar tissue can be seen in Fig. 2.

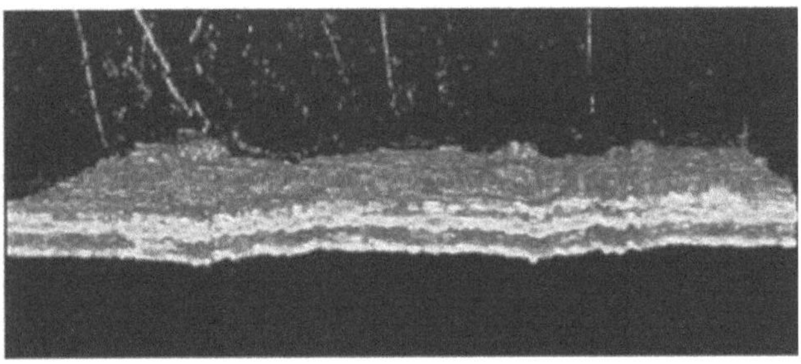

Fig. 2. Visualization of the alveolar tissue through 3D texture slicing volumetric rendering.

Theory of the Mathematical Model

We model both the inhaled air into the alveolar tissue and the blood flowing to get oxygenated as incompressible viscous fluids. As can be seen in Fig. 3a, b, we have two fluid chambers and the tissue domain for diffusion in between them.

The tissue cells where the air diffuses through are modeled as the boundary for those two viscous fluids. However, to simulate the diffusion through the tissue, we needed to define a bounded domain for it. In Fig. 3b, the complex surface boundaries for the tissue domain are shown.

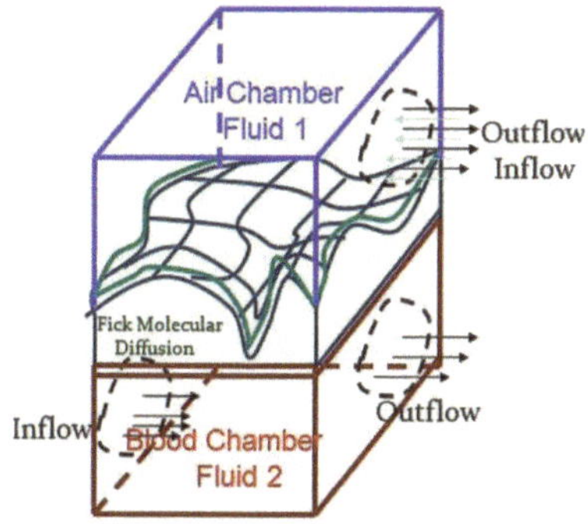

Fig. 3a. Illustration of the mathematical model with two fluid chambers and the alveolar tissue in between: a 3D representation.

Fig. 3b. Illustration of the mathematical model with two fluid chambers and complex surface boundaries for the alveolar tissue: a 2D representation.

Air and Blood as Incompressible Fluids

Generally all fluids are governed by Navier–Stokes equations. These equations represent the state of the fluids and how this state evolves with respect to time and space. Navier–Stokes equations can be obtained from basic Newton laws, conservation of mass, and conservation of momentum. A detailed derivation of Navier–Stokes can be obtained from Griebal et al. [14] and Chorin and Marsden [15]. In what follows, we briefly summarize the governing equations for incompressible fluids.

$$\frac{d}{dt}\int_{\Omega_t} \rho(\mathbf{x},t)d\mathbf{x} = 0 \tag{1}$$

Equation (1) states that the derivative of the mass with respect to time should vanish to conserve mass. In (1), $\rho\ (x,t)$ represents the density of the fluid, and Ω_t denotes the domain of the fluid. By applying Reynolds' transport theorem, with $\boldsymbol{u}$ representing the velocity field, we get the following equation:

$$\frac{\partial \rho(\mathbf{x},t)}{\partial t} + \nabla \cdot (\rho \mathbf{u}(\mathbf{x},t)) = 0 \tag{2}$$

For incompressible fluids, the first term vanishes; hence, the conservation of mass states that incompressible fluids should have a non-divergent velocity field, which is given in (3) as

$$\nabla \cdot \mathbf{u}(\mathbf{x},t) = 0 \tag{3}$$

The momentum of the fluid can be defined over the fluid domain Ω_t as follows:

$$\mathbf{m}(\mathbf{x},t) = \int_{\Omega_t} \rho(\mathbf{x},t)\mathbf{u}(\mathbf{x},t)d\mathbf{x} \tag{4}$$

The second Newton law states that the momentum of the fluid should be preserved. In other words, the rate of change of the momentum with respect to time should be equal to the summation of the forces applying on the fluid expressed as

$$\frac{\partial}{\partial t}\int_{\Omega_t} \rho(\mathbf{x},t)\mathbf{u}(\mathbf{x},t)d\mathbf{x} = \int_{\Omega_t} \rho(\mathbf{x},t)\mathbf{g}(\mathbf{x},t)d\mathbf{x} + \int_{\partial\Omega_t} \boldsymbol{\sigma}(\mathbf{x},t)\mathbf{n}ds \tag{5}$$

Equation (5) simply states that the change in the momentum is a sum of the body force density and the internal surface forces summation. These internal forces can be a combination of a surface force, such as pressure, and stresses, such as viscosity drag forces. In (5), $\boldsymbol{\sigma}\ (x,t)$ represents the stress tensor. After applying the transport theorem, the product rule, and the divergence theorem, and incorporating the incompressible fluids non-divergent velocity field provided by (3) into (5), we can write

$$\frac{\partial}{\partial t}\mathbf{u}(\mathbf{x},t) + \mathbf{u}(\mathbf{x},t)\cdot\nabla\mathbf{u}(\mathbf{x},t) + \frac{1}{\rho}\nabla p(\mathbf{x},t) = \frac{\mu}{\rho}\nabla^2\mathbf{u}(\mathbf{x},t) + \mathbf{g}(\mathbf{x},t) \tag{6}$$

In (6), $p(x,t)$ represents the pressure, and μ represents the dynamic viscosity of the fluid. In order to model the fluid mass propagation and the rate of change of density field, we follow Stam's lead [4]. We derive the rate of change of density relating the average mass fluxes in terms of the diffusion and advective drag effect of the velocity as

$$\frac{\partial}{\partial t}\rho(\mathbf{x},t)+\mathbf{u}(\mathbf{x},t)\cdot\nabla\rho(\mathbf{x},t)=\kappa\nabla^2\rho(\mathbf{x},t)+s(\mathbf{x},t) \tag{7}$$

In (7), κ is the diffusion constant. The rate of change of density at a fluid cell depends upon the diffusion and the current velocity moving through the fluid cell. The $s(x,t)$ term is incorporated into the linear advection equation to account for the sources added into the simulation apart from the inflow boundary. Equations (6) and (7) together with (3) form the basis for the simulation of the air and blood flow in our alveolar fluid dynamics model. In [16], the authors develop a solution procedure for convective diffusion, similar to (7), and provide example solutions for shear and elliptic flow.

Fick's Law of Diffusion and Darcy's Law for Diffusion Through Alveolar Tissue

The diffusion of a fluid is governed by Fick's law that predicts how the diffusion changes the density field with respect to time and space. Fick's law is given as

$$\frac{\partial}{\partial t}\rho(\mathbf{x},t)=\kappa\nabla^2\rho(\mathbf{x},t) \tag{8}$$

which is used to model the diffusion through the alveolar tissue. The alveolar tissue is basically modeled as a fluid where the velocity field is always zero, which is a direct implication of Fick's law and is satisfied on the permeable boundaries through a free-slip boundary condition for velocity.

The amount of air that is coming into the tissue from the air chamber needs to be determined initially for the diffusion through the tissue chamber. We model the air–tissue and tissue–blood boundaries as porous walls of the alveolar tissue, permitting the air to pass through. The average mass flux through these walls is given by Darcy's law that relates the average mass flux to permeability and pressure gradient, as

$$\psi=-\frac{\tau}{\mu}\nabla p(\mathbf{x},t) \tag{9}$$

In (9), τ represents permeability of the alveolar tissue; ψ gives the average mass flux.

Algorithms, Numerical Techniques, and Details of Implementation

In order to solve the partial differential equations (PDEs) (6), (7), (8) and (3) we make use of the Helmholtz Hodge decomposition theorem [14, 15], which is given as

$$\mathbf{u}(\mathbf{x},t)=\mathbf{u}_{\mathbf{df}}(\mathbf{x},t)+\nabla p(\mathbf{x},t) \tag{10}$$

Equation (10) tells us that a vector field, $\boldsymbol{u}$, in fluid velocity field, can be decomposed into a divergent-free (mass conserving), $\boldsymbol{u}_{df}$, and curl-free (irrotational), $\nabla \mathrm{p}$, vector fields. The irrotational part can be represented as a gradient of a potential field, which is the pressure. By taking the divergence of (10), we can write the Poisson pressure equation as

$$\nabla \cdot \mathbf{u}(\mathbf{x},t)=\nabla^2 p(\mathbf{x},t) \tag{11}$$

In order to solve these PDEs given in (3), (6), (7), (8), (9), (11) numerically, the domain of the fluids needs to be discretized. We use collated uniform grids, although more accurate discretization methods exist in the literature, i.e., staggered grids, Marker and Cell (MAC) [17, 18], boundary-fitted grids [19]. We chose collocated grids given their ease of implementation on GPUs with floating point textures. As for the boundary conditions for the fluids, we have used the Neumann boundary condition for pressure and density, which states that there is no change in the density or pressure along the normal direction at the boundary. We have used the free-slip boundary condition for velocity to further guarantee that no flow penetrates into the boundary. We obtained the complex geometry of the boundary through a segmentation carried out on the OCM-imaged alveolar tissue. We flagged the cells in the domain as air, blood, tissue cells, and boundary cells wrapping the fluid chambers. The air–tissue and tissue–blood boundaries were further flagged as permeable boundaries. Figure 4 shows the gridding, boundary, and tissue marking.

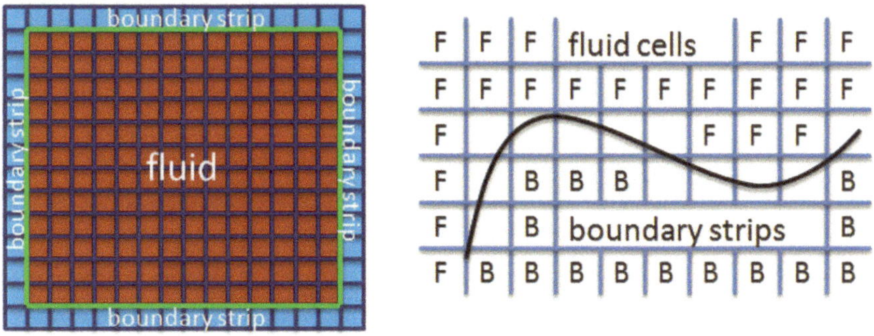

Fig. 4. Uniform grids, boundary strips, marking of cells as boundary and fluid, i.e., tissue, air, or blood.

In order to effectively represent the boundary and provide accurate simulations, we have created 3×3 marking kernels in the marking of the boundary cells to overcome the sharp edges in the boundary which cause discontinuities throughout the boundary when skipped. To simulate the permeability, we have deducted the average mass flux calculated by (9) from the neighboring air cells to the air–tissue boundary and added this amount of mass flux into the tissue as new mass sources in each frame of the simulation. We have maintained the Neumann boundary condition for density while still simulating the fluid diffusing through the membrane.

In order to satisfy the stability requirements, we simulate the elliptic and parabolic PDEs with implicit methods instead of explicit methods. This approach also helps the real-time performance, because we can time step bigger intervals. An excellent coverage of the solution of the PDEs with explicit and implicit methods can be found in [20]. To implement the implicit methods for the elliptic PDEs, we make use of Jacobi iterative technique for the solution of $\mathbf{Ax} = \mathbf{b}$, where $\mathbf{A}$ is a tridiagonal with fringes matrix, $\mathbf{x}$ contains the next time step density or pressure values; $\mathbf{b}$ contains the current simulation state. For the advective terms appearing in (6) and (7) we have used Stam's stable semi-Lagrangian method [4]. All the differential terms are approximated by the central difference method.

In Fig. 5, the steps of the algorithm to simulate the incompressible viscous fluids are illustrated. At each time step, the nonlinear momentum equation is iterated with implicit viscous diffusion and semi-Lagrangian advection steps. Then the values obtained for the flow are immediately used to move the densities around along with the solution of the linear convective diffusion given in (7). In the tissue domain, only the Fick law's diffusive equation (8) is solved, since there is no flow inside the tissue. The source term in (7) represents the amount of density coming into the tissue from the air chamber along with the air diffusing from the tissue to the blood chamber. As can be seen in Fig. 5, at each time step, the results are visualized through an orthographic projection and an appropriate color coding embedded.

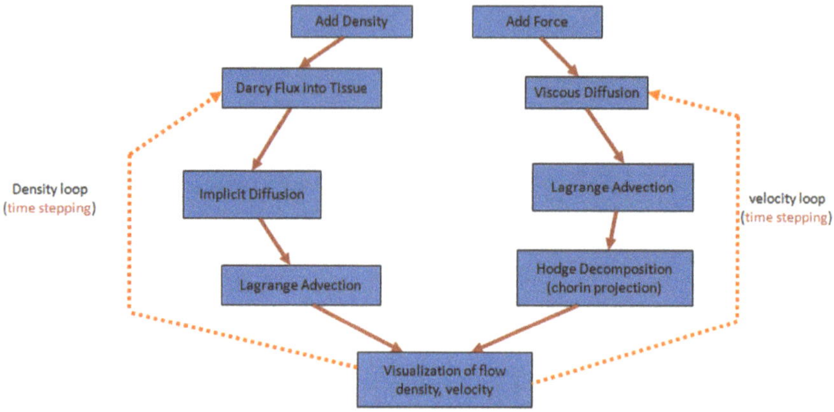

Fig. 5. The steps of the algorithm for incompressible viscous flow and diffusion through the alveolar tissue.

Results

We have implemented the simulator and run the tests with a Dell XPS710 system with Intel Core 2 CPU at 2.4 GHz, 2 GB RAM, and the NVIDIA 8800 GTX graphics card. All the parts of the simulator were implemented with DirectX 10 Graphics API with floating point textures and render target principles in order to run the simulator completely on the GPU. Since the tissue engineers have to use a water-like liquid instead of blood in their growing tissue cultures, we have made the simulations with the water parameters instead of blood. In Table 1, all the parameters used in the simulations are shown.

Table 1. Parameters used in the alveolar tissue simulation [14, 21].

Chambers	Length [μm]	Depth [μm]	Grid spacing dx [μm]	Density ρ[kg/m^3]	Dynamic viscosity μ [kg/ms]	Diffusion constant κ [m^2/s]	Permeability τ, darcy [10^{-12}m^2]	Partial pressure, O_2 [mmHg]
Air	200	125	0.26	1.165	1.86×10^{-5}	1.88×10^{-5}	–	100
Tissue	200	25	0.26	–	–	1.0×10^{-5}	0.062	varied
Water	200	50	0.26	995.7	7.98×10^{-4}	1.9×10^{-9}	–	40

The simulator currently runs only in 2D, and the velocity and the density fields are visualized at the end of each time step. The green isosurfaces show the magnitude of the velocity field in downward direction (+z), vertical to the tissue surface, and the red isosurfaces show magnitude of the velocity field in the axial direction

(+x). The combination of the colors, such as orange and yellow, shows that the velocity field has two components and both of them are non-zero. The density field is represented as white inside the air chamber, green inside the porous alveoli tissue, and red inside the blood chamber. The density and the velocity sources are added into the simulation by clicking the mouse, since we have not yet estimated alveolar ventilation flow into the air chamber. Results are shown in Figs. 6 and 7.

Conclusion

In this chapter, we have set up a physics-based mathematical model to simulate the diffusion of the air through the permeable alveolar tissue constructs imaged with an OCM system developed in our laboratory. The physics-based mathematical model incorporates the physical properties of the alveolar tissues. Our model combined with a physically correct 4DCT registration algorithm, which provides the input air flow (i.e., regional ventilation), will be used in future work to predict parameters of the alveolar tissues. For instance, the inverse analysis of Darcy's law will be performed to get the correct permeability values of the porous tissue. We have developed the simulator on GPUs to be able to provide the real-time simulations of the mathematical model. Our GPU simulations are one of the first of its kind to provide the flow across a porous medium with permeable boundaries, which has not been implemented on GPUs to-date. Furthermore, our model uses the exact dimension of the engineered alveolar tissue and takes into account the complex geometry of the tissue by OCM optical imaging.

Acknowledgments The authors of this work are grateful to the LINK Foundation and the US Army Medical Research and Material Command for supporting this work. We would like to thank Vineet Goel, Sumanta Pattanaik, and Olusegun Ilegbusi for stimulating discussions on GPUs, computer graphics, and computational fluid dynamics, respectively, and Ayesha Mahmood from VaxDesign for providing the engineered tissue samples.

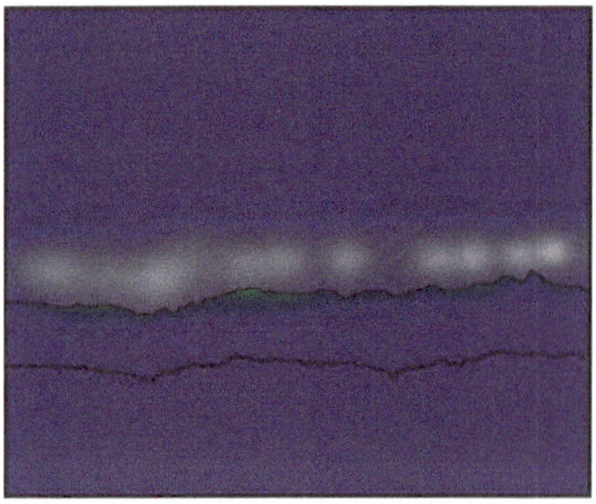

(a) The initial density field in the air chamber at 1 s.

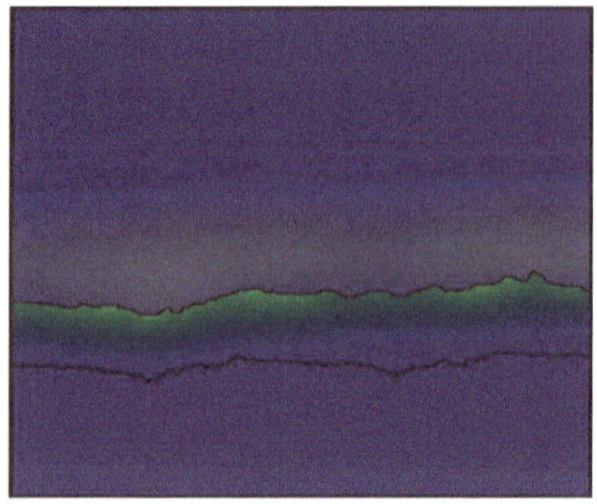

(b) Diffusion starts into the tissue at 5 s.

(c) Diffusion reaches the blood chamber at 15 s.

(d) The tissue is oxygenated at 2 min.

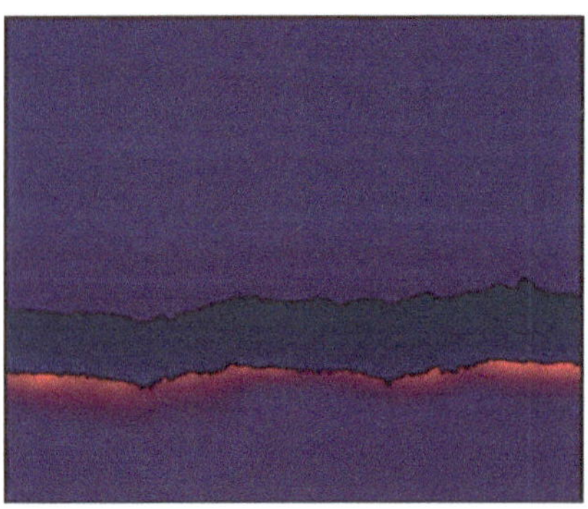

(e) Low diffusivity in the blood chamber, at 4 min

(f) All the gas has diffused into the blood chamber at 10 min.

Fig. 6. The diffusion of the air through the alveolar tissue with complex surface boundary shown as the evolution of the density profile.

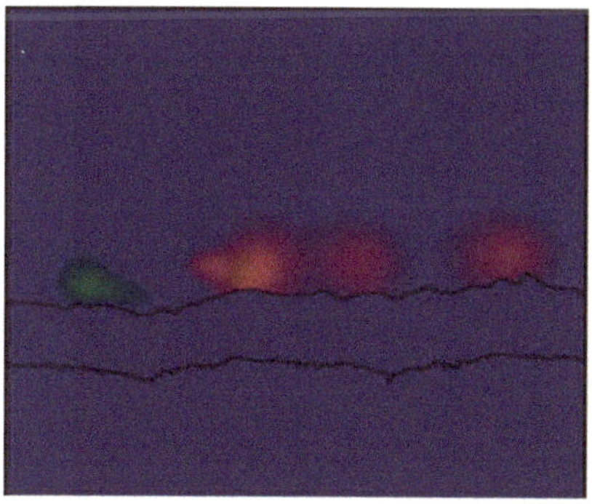

(a) The initial air flow in the air chamber at 1 s.

(b) The flows in both chambers with different viscosities at 10 s.

(c) Viscous diffusion and self-advecting flows in both chambers at 90 s.

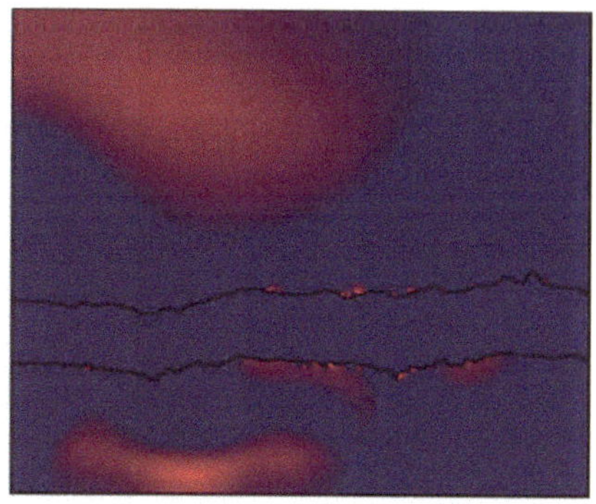

(d) The divergence of flows used in the velocity correction at 2 min.

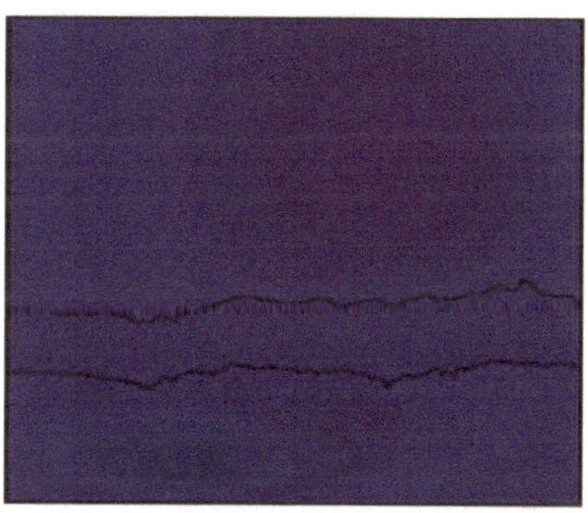

(e) The flows are affected by viscosity, flow profiles at 4 min.

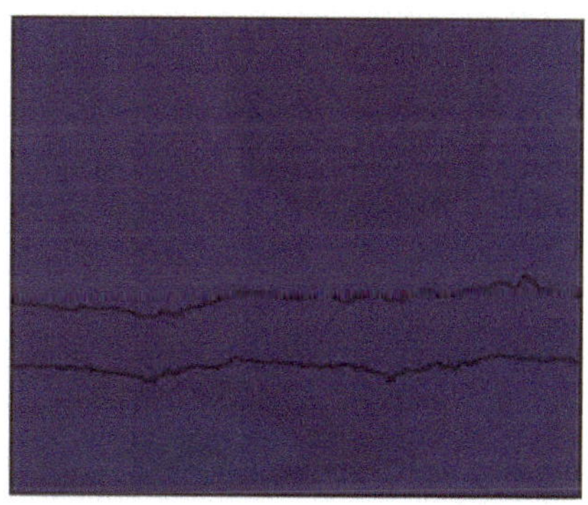

(f) The frictional effect of the viscosity diminishes the flows in both chambers at 10 min.

Fig. 7. The evolution of incompressible viscous flow profiles in the air and the blood chambers.

References

1. Santhanam A, Imielinska C, Davenport P et al. (2008). Modeling and simulation of real-time 3D lung dynamics. Trans Inf Technol Biomed 12(2):257–270
2. Kaya I, Santhanam A, Imielinska C, Rolland JP (2007) Modeling air-flow in the tracheobronchial tree using computational fluid dynamics. In: Proceedings of 2007 MICCAI workshop on Computational Biomechanics, pp. 142–151
3. Rolland J, Lee K et al. (2008) Collaborative engineering: 3-D optical imaging and gas exchange simulation of in-vitro alveolar constructs. Stud Health Technol Inform 132:426–132
4. Stam J (1999) Stable fluids. In: Proceedings of 1999 ACM SIGGRAPH, pp. 121–128
5. Fedkiw R, Stam J, Jensen HW (2001) Visual simulation of smoke. In: Proceedings of 2001 ACM SIGGRAPH, pp. 23–30
6. Courant R, Issacson E, Rees M (1952) On the solution of nonlinear hyperbolic differential equations by finite differences. Comm Pure Appl Math 5:243–255
7. Wendt JD, Baxter W, Oguz I et al. (2007) Finite volume flow simulations on arbitrary domains. J Graph Models 69(1):19–32
8. Batty C, Bertails F, Bridson R (2007) A Fast Variational Framework for Accurate Solid-fluid Coupling. ACM Trans Graph. doi:10.1145/1276377.1276502
9. Kruger J, Kipper P et al. (2005) A particle system for interactive visualization of 3D flows. IEEE Trans Vis Comput Graph 11(6):744–756
10. Harris M, Baxter W et al. (2003) Simulation of cloud dynamics on graphics hardware. In: Proceeding of 2003 ACM SIGGRAPH workshop on Graphics Hardware, pp. 92–101
11. Izatt J, Hee MR, Owen GM, Swanson EA, Fujimoto JG (1994) Optical coherence microscopy in scattering media. Opt Lett 19(8):590–592
12. Murali S, Thompson KP, Rolland JP (2009) Three-dimensional adaptive microscopy using embedded liquid lens. Opt Lett 34(2):145–147
13. Engel K, Hadwiger M et al. (2006) Real-Time Volume Graphics. A. K. Peters Ltd, Wellesley, USA
14. Griebal M, Dornseifer T, NeunHoeffer T (1998) Numerical Simulation in Fluid Dynamics – A practical Introduction. Society for Industrial and Applied Mathematics (SIAM), Philadelphia, USA
15. Chorin AJ, Marsden JE (2000) A Mathematical Introduction to Fluid Mechanics (4th edition). Springer, New York
16. Bernard PS (1990) Convective diffusion in two-dimensional incompressible linear flow. SIAM Rev 32(4):660–666
17. Harlow F, Welsch J (1965) Numerical calculation of time-dependent viscous incompressible flow of fluid with free surface. Int Phys Fluids 8:2182–2189
18. Tome MF, MacKee S (1994) A computational marker and cell method for free surface flows in general domains. J Comput Phys 110:171–186
19. Engel M, Griebel M (2006) Flow simulation on moving boundary-fitted grids and application to fluid-structure interaction problems. Int J Numer Meth Fluids 50:437–468
20. Press WH, Flannery BP, Teukolsky SA et al. (1992) Numerical Recipes in C (2nd edition). Cambridge University Press, New York
21. West JB, (2005) Respiratory Physiology – The Essentials. Lippincott Williams & Wilkins, Baltimore

Part III

Medical Analysis and Knowledge Management

Estimating Hip Joint Contact Pressure from Geometric Features

Ehsan Arbabi[1], Salman Chegini[2], Ronan Boulic[3], Stephen J Ferguson[4], and Daniel Thalmann[5]

Abstract Hip mechanical simulation for estimating pressures within the soft tissues during the loads and motions of daily activities is a common approach to investigate hip joint pathology. Many computational approaches estimate the pressure and contact pressures via finite element methods (FEM) by using 3D meshes of the tissues. Although this type of simulation can provide a good evaluation of hip problems, the process may be very time consuming and unsuitable for fast medical hip simulations. In this chapter, a statistical model is proposed for estimating hip pressures during its movement, by using the geometrical features extracted from 3D meshes of different hip models. The method is tested by examining 25 different hip models during a frequent daily activity.

Introduction

Nowadays, computer-aided simulations of human joints can help medical doctors and surgeons to achieve a faster and more precise diagnosis or surgical plan [1, 4, 8, 10]. Hip problems such as pain related to the acetabular rim and labrum have received increased attention in the orthopedic literature [7]. Two different types of femoroacetabular impingements have been considered. Pincer impingement occurs between the anterior wall or acetabular labrum and the femoral head.

[1] Virtual Reality Lab., École Polytechnique Fédérale de Lausanne, Switzerland
ehsan.arbabi@epfl.ch

[2] MEM Research Center, University of Bern, Switzerland
salman.chegini@artorg.unibe.ch

[3] Virtual Reality Lab., École Polytechnique Fédérale de Lausanne, Switzerland
ronan.boulic@epfl.ch

[4] MEM Research Center, University of Bern, Switzerland
stephen.ferguson@artorg.unibe.ch

[5] Virtual Reality Lab., École Polytechnique Fédérale de Lausanne, Switzerland
daniel.thalmann@epfl.ch

N. Magnenat-Thalmann et al. (eds.), *Recent Advances in the 3D Physiological Human*,
DOI 10.1007/978-1-84882-565-9_11,

Cam impingement occurs when the femoral head–neck junction has an abnormally large radius resulting in insufficient offset [12].

The mechanical contributions of the cartilage layers inside a joint include articulation and load transfer, during which two cartilage layers exert contact pressure on each other. This pressure has been a notable parameter to evaluate the physical conditions inside the hip joint. High pressures are shown to be in association with soft tissue damage within the hip joint. Therefore, in vivo or in vitro measurements of contact pressure within the hip joint have been performed parallel to the development of predictive tools [9].

Many computational approaches estimate the pressure and contact pressures via finite element methods (FEM) [16], during the loads and motions of daily activities, by using 3D meshes of the tissues [4]. Although this type of simulation can provide a good evaluation of hip problems, the process may be very time consuming and unsuitable for fast medical hip simulations. Also, lack of the details related to the joint movement can completely prevent the simulation to be performed.

In this chapter, we propose a statistical model for estimating hip contact pressures during its movement. The estimation is done by extracting and evaluating the geometrical information of the target hip and some other already investigated hip models. The method does not require any details about the joint movement, and having 3D meshes of the joints is sufficient. The method has also been examined on 25 different hip models, and the estimated pressures were compared to the FEM-based calculated pressures.

Method

We have some already investigated hip models (training models), and a non-investigated hip model (target model). The contact pressures of the training models during a specific movement are available. The goal is to estimate the pressures of the target model during the same movement, by using the key geometrical features of the training models. The main steps of the method can be listed as follows:

1. geometric feature extraction and reduction from the training and target models;
2. finding the closest training models to the target model, based on the features;
3. pressure estimation based on the closest training models.

For simplicity, let us consider that we have **'m'** (>0) number of training hip models and one target hip model. For each training hip model, **'p'** number of values is available, where each value represents the maximum contact pressure of the hip model at a hip position (i.e., step of rotation during hip movement):

$$\{\text{Pressure values of the } i\text{th training set}\} = \{\{\text{Pr}_j\}_i\}$$

Pressure values of the target set = $\{Pr_j\}_{target}$, i=1 to **m** and j=1 to **p**.

For estimating the pressure of the target model, we extract '**f**' number of features from all the hip models (training and target sets):

Training features = {Feature array of the ith training set}={ $\{F_l\}_i$ }

Target features = Feature array of target set = $\{F_l\}_{target}$, i=1 to **m** and l=1 to **f**.

Geometrical Feature Extraction

Since hip 3D meshes may have different resolutions, we cannot use the raw 3D coordinate of the hip vertices as input features. Instead, we sample the tissue meshes with constant sampling steps for all the hip models. Due to spherical shape of hip joint, the sampling is performed in the spherical coordinate system, rather than Cartesian coordinate system. For each sample point the distance between the sample point and the joint center (R) is stored as a feature.

Tissue sampling provides a large number of R values as feature. Processing this large number of data can slow down the total pressure estimation process and consequently make the process less efficient. The principal component analysis (PCA) discriminates directions with the largest variance in a data set for identifying the most representative features [11]. We apply PCA on the normalized features (R's) and keep '**e**' number of the most effective output features (Effective features = {PCA-based effective features of the ith training set} = $\{\{EF_k\}_i\}$, i=1 to **m** and k=1 to **e**). The most effective features of the target set is also found by using the same PCA parameters calculated for training sets (Target effective features=$\{EF_k\}_{target}$, k=1 to **e**).

In order to find the pressure-correlated features in each step of hip movement, we calculate the statistical correlation coefficient between each effective feature and the contact pressure at a specific hip position (i.e., step of rotation), for all the training hip models. Based on these correlation coefficients, the features can be ranked and **e'** number of the highest ranked ones are considered for later computations. Since the pressures are related to a range of motion rather than just a specific hip position, these correlation coefficients calculations are done for all the hip movement steps:

Correlation coefficients = {Correlation coefficient between pressure at the jth step of hip movement and the kth effective feature, for all training sets} = {Correlation coefficient between $Pr_{j,i=1 \text{ to } \mathbf{m}}$ and $EF_{k,i=1 \text{ to } \mathbf{m}}$} = $\{\{CC_k\}_j\}$, k=1 to **e** and j=1 to **p**.

We also define HCC(**e'**,j) as a set containing the indices of **e'** ($\leq$**e**) number of the effective features with the highest correlation coefficient at the *j*th step of hip movement:

HCC(**e'**,j)={k | CC_{kj} is among the **e'** number of the highest correlation coefficients}.

Finding the Closest Set Per Feature

Among all the training sets we search for the set which has the smallest distance to the target set. The distances are calculated by using the extracted features; thus for each feature (k), we find the training set (i) which returns the smallest distance between $EF_{k,target}$ and all the $EF_{k,i}$:

Closest set corresponding to all the effective features = {Closest set for feature k} = $\{CS_k\}$ = $\{MIN|_i(DIST\langle EF_{k,target}, EF_{k,i}\rangle)\}$, i=1 to **m** and k=1 to **e.**

There are different methods for finding distance between two sets [18]. For finding the smallest distance between $EF_{k,target}$ and all the $EF_{k,i}$'s, two kinds of Euclidean distances are considered: DIST1 and DIST2. $DIST1(B,A)_k$ is Euclidean distance between array A and array B, when the kth elements in both arrays are ignored, which demonstrates how much the difference between sets of A and B is depending on the kth element. $DIST2(B,A)_k$ is the absolute difference between the kth elements in array A and array B. It demonstrates how the kth element of sets of A and B are close to each other. For each feature, we first find four training sets with the smallest DIST1 to the target set and then among these four training sets, we selected the one with smallest DIST2 (four was chosen by trial). At the end, '**e**' number of training sets, corresponding to '**e**' number of effective features, is found (with possible repetition).

Estimating Pressures

The pressure of the target hip at the *j*th step of movement is estimated as the weighted average pressures of the **e'** number of closest training sets to the target set (i.e., CS), corresponding to **e'** number of high correlated effective features. The weights are the correlation coefficients of each effective feature ($\{CC_k\}_j$):

$$Pr_{j,target} = \sum^{k \in HCC(\mathbf{e'},j)} (CC_{k,j} * Pr_{j,i}) / \sum^{k \in HCC(\mathbf{e'},y)} (CC_{k,j}), \qquad \text{where } i = CS_k.$$

Testing the Method on Different Hip Models

Hip Models

The α and CE angles are two parameters which can characterize some hip joint pathologies. The CE angle (Fig. 1) is defined as the angle formed by the perpendicular to the inter-tear drop line and the line passing from the center of the femoral head to the lateral edge of the acetabulum [19].

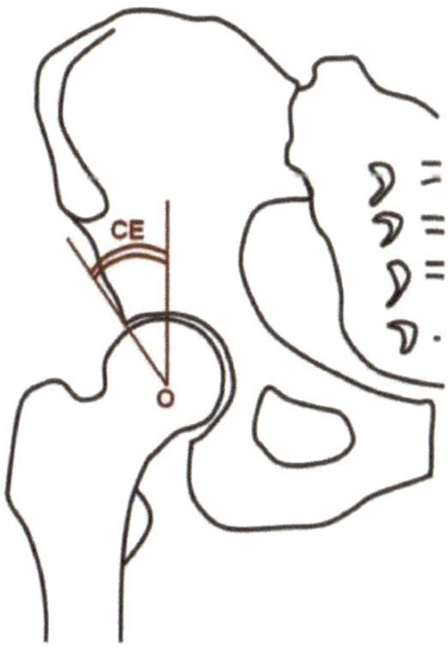

Fig. 1. CE angle in a hip joint [5].

The α angle is measured as the angle between A–O and O–O', where O is the center of the head, O' is the center of the neck at the narrowest point, and A is the anterior point where the femoral head diverges from spherical [15] (Fig. 2).

The hip models were created using CAD software [17], with variety of α and CE angles to cover a wide range of hip geometries. The CE angle values were 0°, 10°, 20°, 30°, 40° and α angles were 40°, 50°, 60°, 70°, 80°. Combination of all possibilities of the α and CE angles produced 25 joints varying from normal (CE=20°, α=40°) to a dysplastic (CE=0°, α=40°), cam (CE=20°, α=80°), and pincer joint (CE=40°, α=40°) (see Fig. 3) [5].

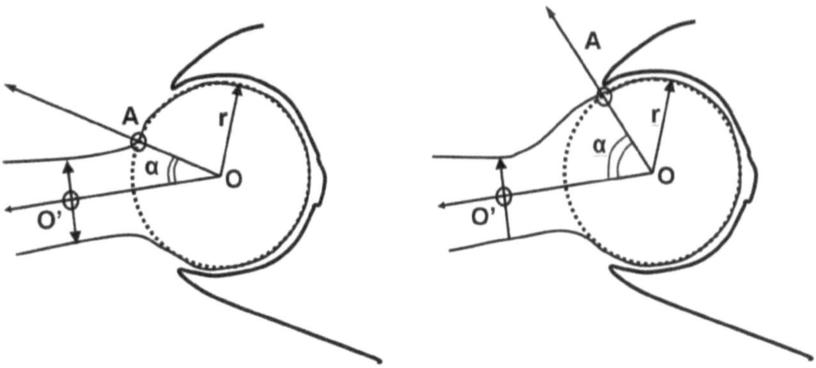

Fig. 2. *Left*: **α** angle in a normal joint. *Right*: **α** angle in a 'cam'-type joint [5].

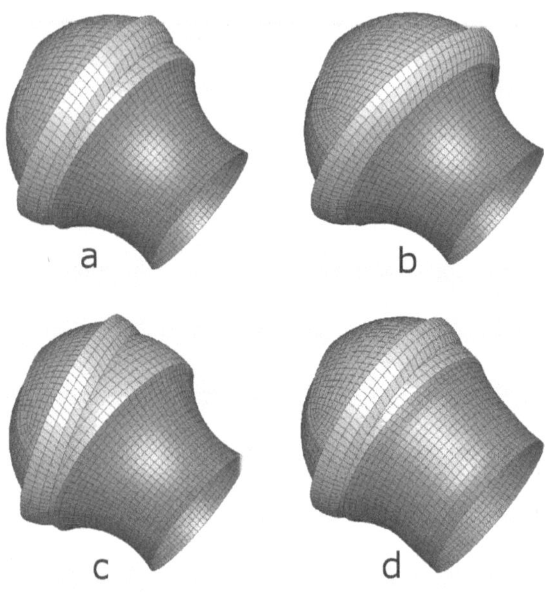

Fig. 3. (**a**) Normal joint, (**b**) pincer joint, (**c**) dysplastic joint, (**d**) cam-type joint [5].

Since α and CE angles are geometrical parameters, they could also be included in the feature lists. In fact, having α and CE angles can increase the accuracy of the estimation as they are correlated to the hip joint pathology. However, calculating α and CE angles in real human hip joint may not be straightforward. Especially for α angle, the hip head has to be modeled as an ideal sphere in order to find the point where femoral head diverges from spherical [15]. Thus, such values can be highly depending on the methods used for their calculations. In order to have a more robust method, we avoided using α and CE angles among our features.

Hip Movement

We wanted to test our method during a medically meaningful hip movement. It has been shown that hip impingement is a kinematical problem associated with large range of motion and is most often observed in young, active patients [13, 3]. Therefore, the motion data for standing-to-sitting were chosen as representative of the frequent daily activities with a relatively large range of motion for comparison [2]. We divided the total movement to 50 continuous steps and estimated the pressure for them.

Pressure Calculation by FEM

These models were imported to finite element software (ABAQUS 6.6, ABAQUS Germany Ltd., Aachen, Germany). The motion data were applied as a prescribed rotational kinematic about the femoral head center. The translation was kept unconstrained and the corresponding joint reaction force vector was simultaneously applied [5]. Cartilage was modeled as a linear elastic material with an elastic modulus of E=12 MPa and Poisson ration of υ=0.45 [14]. The labrum's material properties were E=20 MPa and υ=0.4 [6], and the bony components were assumed to be rigid in comparison to the soft tissues. As outputs of the simulation, contact pressures were reported. In five pathological models, due to the large deformation, the rotation was stopped before being fully accomplished. These models and the percentage of their accomplished rotation are listed in Table 1.

Table 1. The models without full range of rotation.

α angle	CE angle	Percentage of the accomplished rotation (%)
80°	40°	50
80°	30°	71
80°	20°	91
70°	40°	83
60°	40°	87

Pressure Estimation by Using the Proposed Method

If we consider the total number of hip models as '**m**,' we trained and tested the method '**m**' times. Each time, one of the models was chosen as a target model and the rest (**m**–1) were used as the training models. The estimated pressures during the rotation were compared with the FEM-based calculated pressures to evaluate the error. Since five models were not examined during the full rotation (see Table 1), we performed the test for two cases. In the first case, half of the rotation was considered so that all 25 models could be tested. In the second case, the full rotation was considered so that 5 models had to be excluded and 20 models were tested. Totally 1944 features were extracted (**f**=1944), and 10 of them were considered as effective features after applying PCA (**e**=10), which means more than 99% of the features were discarded. We considered **e'** number of the most effective features among these 10 features, based on their correlation coefficients with contact pressures (CC). In order to evaluate the effect of the number of extracted features on the estimated pressures, we considered different values for **e'** (1 to 10) and estimated the pressures for each value of **e'**, separately.

Results

For each test, we computed the error by calculating the difference between the estimated pressure and the FEM-based calculated pressure. The difference was divided by the FEM-based calculated pressure, in order to have the percentage of error (see Fig. 4).

Since the percentage of error was calculated for different steps of rotation, its mean value was calculated during the rotation too. The pressure estimation average error of different hip models (in percentage), when **e'**=10, for the first and the second cases are listed in Tables 2 and 3, respectively.

Table 2. Overall pressure estimation error (percentage) of the hip models for the first case (e'=10).

		CE angle				
		0°	10°	20°	30°	40°
α angle	40°	5.48	1.66	1.63	2.05	4.99
	50°	5.52	2.57	3.21	6.93	17.66
	60°	4.98	1.82	0.72	6.16	–
	70°	5.06	5.74	8.42	8.06	–
	80°	1.24	3.15	–	–	–

Table 3. Overall pressure estimation error (percentage) of the hip models for the second case (e'=10).

		CE angle				
		0°	10°	20°	30°	40°
α angle	40°	4.64	3.75	1.58	1.78	1.85
	50°	4.52	2.79	1.84	1.21	43.59
	60°	4.93	1.22	0.78	1.00	1.69
	70°	4.42	3.00	1.65	1.11	5.34
	80°	8.66	1.71	2.41	0.90	22.17

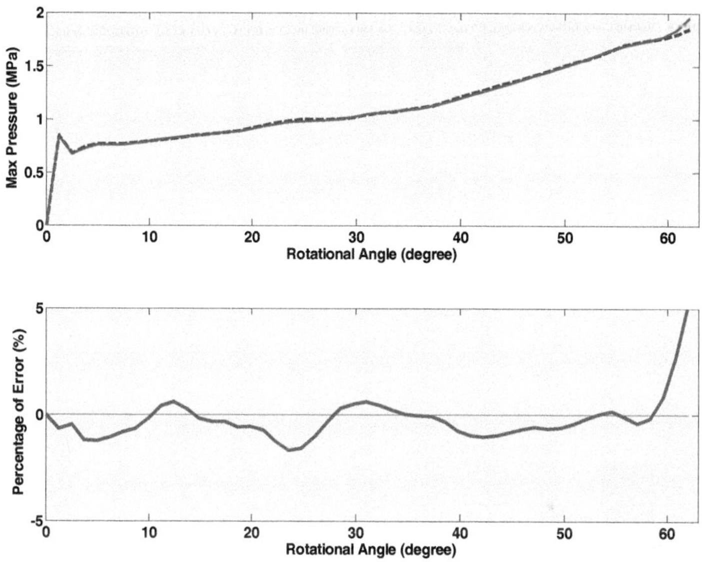

Fig. 4. *Upper*: Estimated maximum contact pressure (*solid*) and FEM-based calculated maximum contact pressure (*dashed*) (α=60°, CE=20°, e'=10). *Bottom*: Percentage of error (*signed*).

For overall evaluation of the method, we calculated the mean value of these average percentages of the error among all the models, for a certain value of **e'** (the number of effective features used for the contact pressure estimation). The overall average errors, related to the different estimations done by considering different values of **e'**, are shown in Figs. 5 and 6.

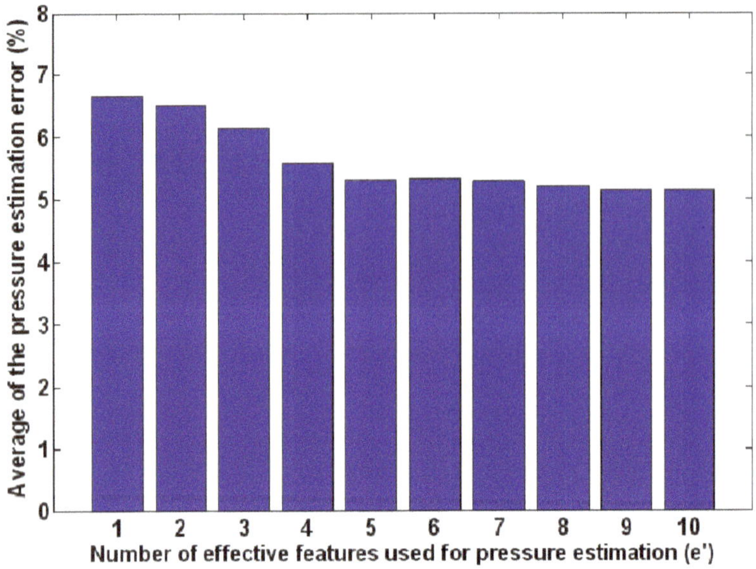

Fig. 5. Average of the pressure estimation error (percentages) for the first case, when e' number of effective feature is used.

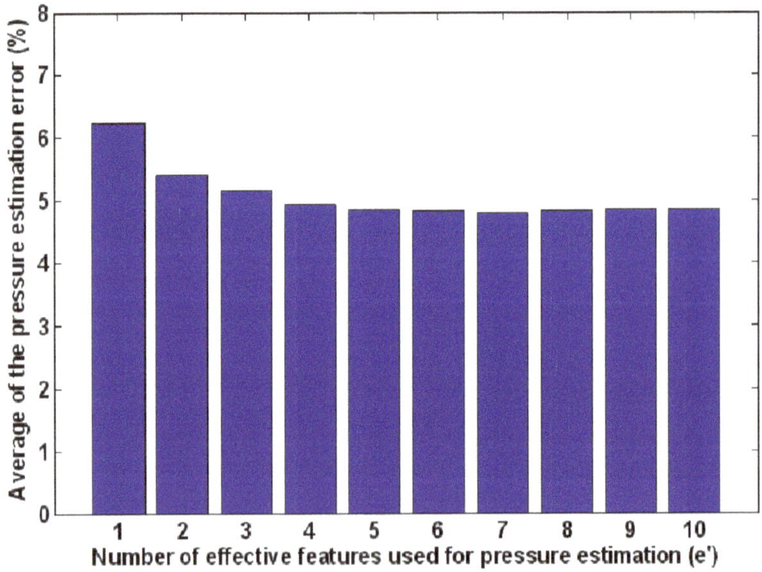

Fig. 6. Average of the pressure estimation error (percentages) for the second case, when e' number of effective feature is used.

Discussion

The results show that the average error of contact pressure estimation is 5% for both cases (Figs. 5 and 6). However, depending on the joint model, the estimation error may vary. For example, as can be seen in Tables 2 and 3, when the target joint is chosen from the bordering joint models (hip models with the maximum/minimum α and CE angles), the estimation error is usually more than the error of the other models. The reason can be due to the lack of enough training models geometrically close to these bordering target models. For example, the estimated contact pressure for the target joint model with $\alpha = 70^\circ$ and $CE = 30^\circ$ is 1.11% in the first case, when the model is in the middle of the training models. But, in the second case the same target model is a bordering model, which causes the estimation error to increase to 8.06%. In fact, if we just consider the non-bordering models as our target models, the estimation error and its standard deviation decrease significantly. For example, in the first case, for **e'**=10, the estimation error is in average 1.62% with a standard deviation of 0.79%, when only the non-bordering target models are considered (the average error is 5.14% with a standard deviation of 9.1%, when the bordering models are also included). This indicates that for having a better estimation it is important to have a wide range of training models in order to ensure that our target models are not bordering.

In Figs. 5 and 6, it can be seen that the estimation error decreases more than 1% when the number of used effective features (**e'**) increases from 1 to 5. That shows having more features can help to have better estimation. However, when the number of used effective features increases more (5 to 10), the error changes less than ~0.2%. So, in general in both cases, using five numbers of effective features can be enough for getting close to the best estimation.

Comparing the first case with the second case shows that the average accuracy of the estimated contact pressures is almost the same. In fact, the average accuracy in the first case is just about 0.6% less than the accuracy in the second case.

Conclusion

In this chapter we proposed a method for estimating contact pressures of the human joints during their movement. The estimation was done based on training a system by the geometrical features of some already investigated joint models. The geometrical features were extracted from 3D meshes of the joints at their default posture. The method was later tested on different pathological hip joints during standing-to-sitting (as an example), and the estimated contact pressures were compared with FEM-based contact pressures. The comparison showed that the accuracy of the method was in average about 95%.

The proposed method suggests a new faster strategy for estimating the contact pressure needed in different biomedical applications, compared to the mechanical models such as FEM (the total processing time in the proposed method was less than 2 s even when all the 10 features were used, by using CPU: Xeon-3.4 GHz and RAM: 2 GB). In addition to the speed, the method does not need the movement details. Such independence from the movement information can be useful when the movement details are either missing or difficult to be mathematically characterized in a simulation.

As a future work, the method can be tested for the other human joints too. Also the method can be improved by evaluating other geometrical features.

Acknowledgments This research has been supported by the NCCR Co-Me (Computer-Aided and Image-Guided Medical Interventions) of the Swiss National Science Foundation.

References

1. Arbabi E, Boulic R, Thalmann D (2007) A Fast Method for Finding Range of Motion in the Human Joints. In: Proc of Int Conf of the IEEE Eng in Medicine and Biology Society, pp. 5079–5082
2. Bergmann G, Deuretzbacher G, Heller M, Graichen F, Rohlmann A, Strauss J, Duda GN (2001) Hip contact forces and gait patterns from routine activities. J Biomech 34:859–871
3. Bizzini M, Notzli HP, Maffiuletti NA (2007) Femoroacetabular impingement in professional ice hockey players: A case series of 5 athletes after open surgical decompression of the hip. Am J Sports Med 35(11):1955–1959
4. Chegini S, Beck M, Ferguson SJ (2006) Femoro Acetabular Impingement as a Possible Initiator of Cartilage Degeneration. In: Proc of Int Symp on Comp Meth in Biomech & Biomed Eng, pp. 705–710
5. Chegini S, Beck M, Ferguson SJ (2008) The effects of impingement and dysplasia on stress distributions in the hip joint during sitting and walking: A finite element analysis. J Orthop Res 27(2):195–201
6. Ferguson SJ, Bryant JT, Ito K (2001) The material properties of the bovine acetabular labrum. J Orthop Res 19:887–896
7. Genoud P, Sadri H, Dora C, Bidaut L, Ganz R, Hoffmeyer P (2000) The hip joint range of motion: A cadaveric study. In: Conf of the European Society of Biomechanics
8. Gilles B, Moccozet L, Magnenat-Thalmann N (2006) Anatomical Modelling of the Musculoskeletal System from MRI. In: MICCAI, pp. 289–296
9. Hodge WA, Fijan RS, Carlson KL, Burgess RG, Harris WH, Mann RW (1986) Contact pressures in the human hip joint measured in vivo. In: Proc Natl Acad Sci (83), pp. 2879–2883
10. Kang M, Sadri H, Moccozet L, Magnenat-Thalmann N (2003) Hip Joint Modeling for the Control of the Joint Center and the Range of Motions. In: Proc of IFAC Symp on Modeling and Control in Biomedical Systems, pp. 20–24
11. Malhi A, Gao RX (2004) PCA-based feature selection scheme for machine defect classification. IEEE Trans Instrum Meas 53(6):1517–1525
12. Mardones RM, Gonzalez C, Chen Q, Zobitz M, Kaufman KR, Trousdale RT (2005) Surgical treatment of femoroacetabular impingement: Evaluation of the effect of the size of the resection. J Bone Joint Surg 87-A:273–279
13. Miller EH, Schneider HJ, Bronson JL, McLain D (1975). A new consideration in athletic injuries. The classical ballet dancer. Clin Orthop Relat Res 111:181–191

14. Moglo KE, Shirazi-Adl A (2003) On the coupling between anterior and posterior cruciate ligaments, and knee joint response under anterior femoral drawer in flexion: A finite element study. Clin Biomech 18:751–759
15. Notzli HP, Wyss TF, Stoecklin CH, Schmid MR, Treiber K, Hodler J (2002) The contour of the femoral head-neck junction as a predictor for the risk of anterior impingement. J Bone Joint Surg Br 84:556–560
16. Russell ME, Shivanna KH, Grosland NM, Pedersen DR (2006) Cartilage contact pressure elevations in dysplastic hips: a chronic overload model. J Orthop Surg 1(6):1–11
17. Solidworks 2005, Solidworks Corp., Boston, MA, USA
18. Theodoridis S, Koutroumbas K (2003) Pattern Recognition. Academic Press, CA
19. Wiberg G (1939) Studies on dysplastic acetabular and congenital subluxation of the hip joint: With special reference to the complication of osteo-arthritis. Acta Chir Scand 58:7–38

Rapid Impingement Detection System with Uniform Sampling for Ball-and-Socket Joint

Ding Cai[1], Won-Sook Lee[2], Chris Joslin[3], and Paul Beaulé[4]

Abstract Detecting the position and the level of joint impingement is often a key to computer-aided surgical plan to normalize joint kinematics. So far most of the current impingement detection methods for ball-and-socket joint are not efficient or only report a few collided points as the detection results. In this chapter, we present a novel real-time impingement detection system with rapid memory-efficient uniform sampling and surface-to-surface distance measurement feature to estimate the overall impingement. Our system describes near-spherical objects in spherical coordinate system, which reduces the space complexity and the computation costs. The sampling design further reduces the memory cost by generating uniform sampling orientations. The rapid and accurate impingement detection with surface-to-surface distance measurement can provide more realistic detailed information to estimate the overall impingement on the ball-and-socket joint, which is particularly useful for computer-aided surgical plan.

Introduction

"Femoro-Acetabular impingement (FAI) is a common cause of hip pain" due to the abnormal joint shape [1, 2]. "Morphologic abnormalities in the femoral head neck junction or the acetabulum lead to abnormal contact forces occurring during extremes of hip motion" [1, 2]. We are motivated by the medical needs of FAI impingement detection for distance parameters. In the preoperative correction simulation, detecting the impinged regions of the joint and estimating the impingement level are our main goals in order to determine the location and amount of the

[1] School of Information Technology and Engineering, University of Ottawa, ON, Canada
dcai024@uottawa.ca

[2] School of Information Technology and Engineering, University of Ottawa, ON, Canada
wslee@uottawa.ca

[3] School of Information Technology, Carleton University, Ottawa, ON, Canada
chris_joslin@carleton.ca

[4] Division of Orthopaedic Surgery, Faculty of Medicine, University of Ottawa, ON, Canada
pbeaule@Ottawahospital.on.ca

N. Magnenat-Thalmann et al. (eds.), *Recent Advances in the 3D Physiological Human*,
DOI 10.1007/978-1-84882-565-9_12,

correction required. Efficiency, accuracy, and memory cost of the algorithms are considered as the most important specifications. Besides the general detection for collided points, we are interested in the surface-to-surface distance measurement to examine the surfaces between the ball and the socket.

Literature Review

Collision detection (also known as interference detection or impingement detection) is considered as one major area in real-time simulations. In the recent surveys [3, 4], the existing collision detection methods are classified into bounding volume hierarchies (BVH), distance fields, feature-based algorithms, spatial subdivision, etc. BVH-based methods aim on the efficiency and distance fields-based methods provide distance measurement feature.

BVH-based methods have advantages on the efficiency due to the tree data structures (e.g., k-d tree and octree). Various types of bounding volumes are used such as sphere, bounding box, and convex hull. Based on sphere-tree design, the object's surface is partitioned and fitted with spheres as the collision response units [5]. Robust and accurate polygon interference detection library (RAPID) contains routines for building the oriented bounding box tree (OBB-Tree) data structure and fast overlap tests between two OBB-Trees for robust and accurate collision detection, which provides tighter bounding than AABB-trees [6, 7].

Distance field [8]-based methods can also be efficiently processed by testing whether the target is inside or outside the pre-computed distance shield as a reference to the object surface. Rasterization is one technique to generate the multi-dimension distance fields [9, 10]. Usually distance fields are computed only in the Cartesian coordinate system. Besides the expensive time cost to generate the field, the field precision is limited by the memory cost.

In medical area detecting impingement and measuring the stress/strain of the ball-and-socket joint have been researched with wider views. Besides the general methods from computer graphics, many specific approaches such as mathematical methods were proposed. Focusing on the hip impingement issue, Sarni et al. [11] developed a molecular model-based system to measure the stress and strain on the cartilages. Yoshida et al. [12] used discrete element analysis (DEA) to measure the pressure distribution of the hip joint during activities of daily living. Hu et al. [13] developed a spatial-based collision detection algorithm by using a look-up table (LUT) and a linear transform indexing method. Kubiak-Langer et al. [14] used the algorithm to compute the range of motion (ROM) in anterior FAI.

Maciel et al. [15] recently used a spherical sliding method which indexes the triangles for the partitioned space and detects collision based on triangle interpolation. The method is efficient in run-time calculation, but it requires expensive sampling method to non-uniformly examine m sampling rays with the n triangles in $O(n^2)$ in pre-processing. Only the triangles intersected with these sampling rays are sampled and recorded. Such sampling method and the data structure have a triangle capture error issue because of the high possibility of missing triangles (either missing in the gaps of the rays or ignored because the same unit space has already been registered by another triangle). The defects cause a direct impact on the practicability, the accuracy, and the reliability of the system.

Uniform random/non-deterministic sampling is popular in scientific analysis but the irregular point distribution is not suitable for traceable data query and the pattern of point distribution is not repeatable. For uniform deterministic sampling, there are different ways such as subdividing icosahedron [16] and mathematical approaches [17] to equally distribute *n* points on the sphere. They are more used to just save geometry data for sampling and modeling purposes with no concern on the efficiency or the data query requirement for collision detection problems.

Previous Work

Inspired by the spatial collision detection method based on look-up table data structure (LUT) [13], a two-step system was designed. First we go through a preprocessing step (Fig. 1b) to sample the input models' geometry information into a data structure. Then in the simulation step (Fig. 1c) we use the geometry data to detect the impingement. Based on the idea, we developed a *rapid spherical impingement detection* system (*RSID* system) [18] for near-spherical objects. To summarize our previous system it contains two major algorithms:

1. Sampling algorithm: scans and stores the models' geometry data into LUT, which is the major part of the pre-processing step.
2. Impingement detection algorithm: detects the impingements between the models based on the sampled data, which is the major part of the simulation step.

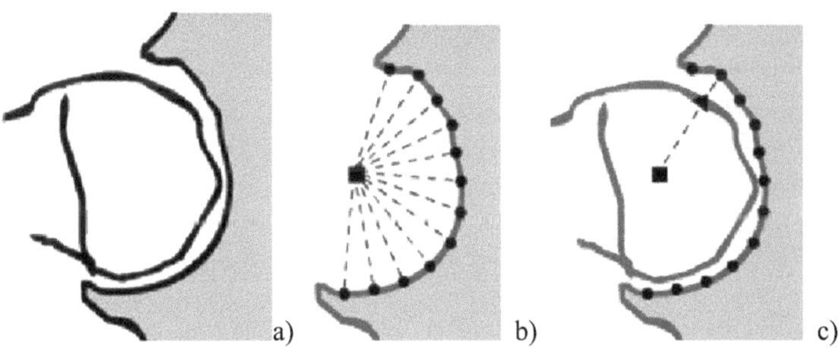

Fig. 1. (**a**) The 2D hip joint example, (**b**) pre-processing step: rectangle is the origin of the spherical coordinate system. *Dots* are the sample points. *Dash lines* are the distances from the sample points to the origin. (**c**) Impingement detection simulation step: the vertex on the ball object is tested against the sampled data in the orientation of that vertex.

Sampling Method in Previous System

In the pre-processing/sampling step of RSID system it samples a near-spherical object based on spherical coordinate system and triangle rasterization. The ball-and-socket joint, such as hip joint, is usually represented by spherical objects in the simulations. In our design we configure the sample points on near-spherical object using their spherical coordinates. In the sampling, the sample points are gained through rapid rasterization in spherical coordinate system. The mesh is first mapped to 2D spherical coordinate grid with the distance data from the origin and then the distances of densely sampled points of the mesh surface are directly interpolated from ones of the triangle vertices through polygon rasterization (Fig. 2).

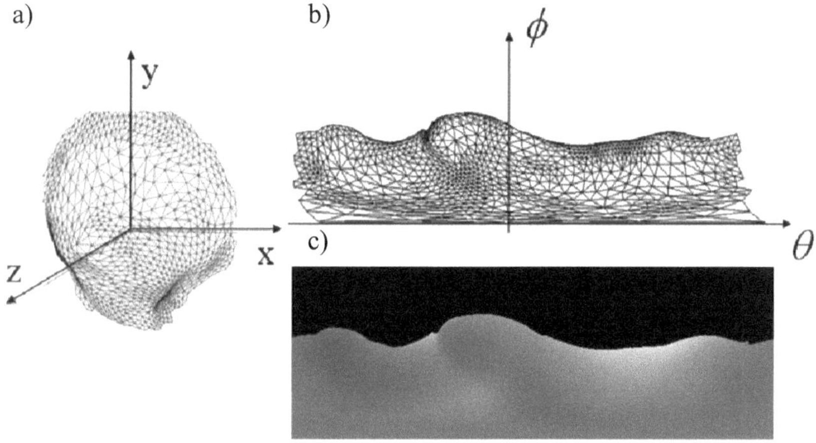

Fig. 2. Rapid sampling: (**a**) socket in the Cartesian coordinates, (**b**) the socket mesh mapped to the 2D spherical coordinate grid, (**c**) the distance data (or distance map) sampled and saved in a 2D LUT corresponding to the spherical coordinates.

We use a large-size LUT, indexed by spherical coordinates, to preserve the geometry distance data gained from the sample points as the proximity information of the object (Eq. 1). The complexity of our sampling algorithm is $O(n)$ to sample a model with n triangles.

$$LUT\ [\phi][\theta] = d \tag{1}$$

where $d \in [0, \infty), \theta \in [0°, 360°)$, and $\phi \in [0°, 180°]$.

Impingement Detection Method in Previous System

In the simulation step of RSID system we use the sampled geometry data stored in the LUT to quickly detect the impingements and locate the impinged points as the common basic results. In the sampling step, we create one LUT for the ball model and one LUT for the socket model. During our ball-and-socket joint simulation, we process impingement detection twice to avoid inter-penetration artifacts: the ball's vertices are tested against the socket's LUT as the first scan and the socket's vertices are tested against the ball's LUT as the second scan. The two impingement detection scans carry the same algorithm so that here we only go through one scan (ball's vertices against socket's LUT) to explain the idea of impingement detection algorithm: in that test, we compute the indices (i.e., ϕ_i and θ_i) of a target vertex v on the ball, check out the socket distance d_i from LUT[ϕ_i][θ_i], and compute the ball vertex's distance d_v to the same spherical coordinate system origin. The impingement detection can be instantly done by comparing the difference between d_i and d_v to find if the vertex is inside or outside of the socket surface. The complexity of our impingement detection algorithm of RSID system is $O(n)$ to scan a model with n vertices against a LUT.

Uniform Spherical Impingement Detection System

The sampling method of our RSID system described in the previous section generates non-uniform sample points because we use regular spherical coordinate grid to define the orientations of the sampling rays in 3D Cartesian space. The non-uniform sampling method is easy to configure and construct but the drawback is that the sampling density is not consistent in all orientations. The sampling density increases at the poles while it decreases at the equator.

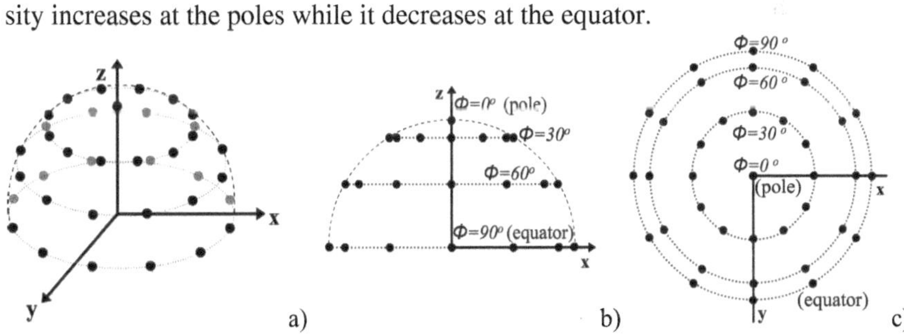

Fig. 3. Non-uniform sampling density on a unit hemisphere at sampling precision=30°: (**a**) perspective view, (**b**) front view, (**c**) top view.

One extreme consequence of the non-uniform sampling is there are many sample points squeezed at the pole while only one sample point is actually needed for the pole location. For example, in Fig. 3 there are 12 sample points corresponding to the 12 θ angles at each ϕ level (sampling precision = 30°). That means at the pole there are also 12 sample points so that the pole will be sampled 12 times. Non-uniform sampling takes some unnecessary computations and memory to store the geometry information.

An alternative impingement detection system was then developed for uniform sampling in 3D Cartesian space, which optimizes the density of the sample points and reduces the memory cost. Like RSID, the new system contains an LUT data structure, a sampling algorithm, and an impingement detection algorithm. This uniform version of RSID is named *uniform spherical impingement detection* system (*USID* system).

The design of uniformly distributing sample points shall still follow the several factors such as

(i) sample points have traceable pattern to be linked to the LUT;
(ii) efficiency and accuracy to access the LUT to check in and check out the sampled distance data;
(iii) easy configuration of multiple sampling resolutions.

Nishio et al. [17] introduced a method to select specific points from a spherical helix curve as the points approximately uniformly distributed on a unit sphere. The spherical helix formula [17] is

$$\theta = 2k\phi \, (0 \leq \phi \leq \pi) \tag{2}$$

where ϕ and θ are spherical coordinates, k is the number of helix turns.

The spherical helix points can be approximately uniformly distributed on the surface of a unit sphere when $k = \sqrt{w}$, where w is the total number of the spherical helix points (Fig. 4). Once k is defined, the helix curve is defined by Eq. 2.

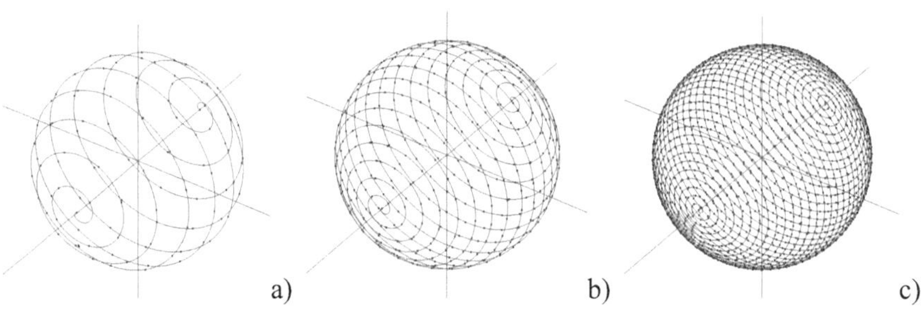

Fig. 4. k-turn spherical helix curve on a unit sphere: (**a**) k=10, w=100; (**b**) k=20, w=400; (**c**) k=40, w=1600.

Look-Up Table Configuration

The new LUT design for uniform sampling is constructed as a 1D array to represent a single *k*-turn spherical helix curve. The size of the LUT is determined by the number of the specified spherical helix points. The data saved in LUT are the distance data sampled in the corresponding orientations. Based on the characteristics of spherical helix, we developed new LUT indexing methods. The LUT check-in and check-out indexing methods will be introduced in the following sections.

Uniform Sampling Method

Specified spherical helix points on a unit sphere are considered another form of sampling orientation configuration (Fig. 5). Equation 2 can specify arbitrary number of spherical helix points, which makes multiple sampling resolutions possible by configuring the curve formula. One example is illustrated in Fig. 4, which shows three different sampling resolutions with three different configurations of the curve formula (Eq. 2).

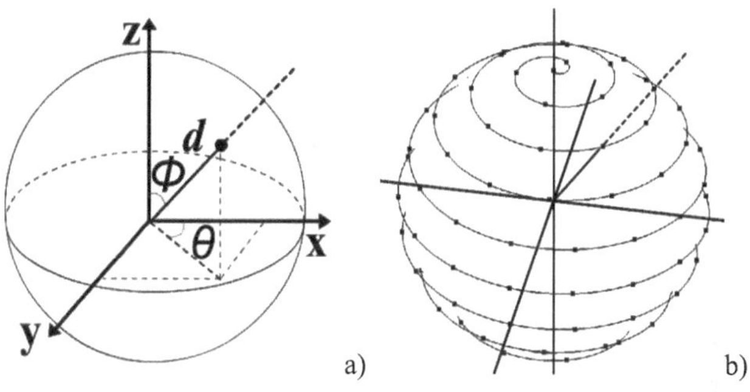

Fig. 5. Define a sampling orientation: (**a**) using spherical coordinates as the common method, (**b**) using specified spherical helix points on a unit sphere in USID system.

We call the points defined by Eq. 2 as *helix orientation points*, which are used to define the sampling rays' orientations. Each *helix orientation point* corresponds to an LUT element. Sliding along the spherical helix curve, one sampling ray is created at one specified *helix orientation point*. The surface distance data are then sampled and checked into the corresponding position of the 1D LUT in order. In this way, the LUT is filled after the spherical helix curve is swept. Originally this is an $O(m*n)$ sampling by testing *m* sampling ray against *n* triangles.

We can take advantage of the rasterization design introduced in our RSID system to accelerate the uniform sampling. First, the non-uniform sampling algorithm of RSID is applied. The distance data are rasterized and fill a 2D temporary LUT as a 2D distance map in $O(n)$ where *n* is the number of the triangles; then, we can sample this 2D distance map with the predefined *helix orientation points*' coordinates and fill our 1D spherical helix-based LUT in $O(n)$ where n is the number of the *helix orientation points*. The 2D temporary LUT can be reused for multiple objects in the scene. It is released from the memory after the sampling step. Since in this case the samplings are based on interpolation, the accuracy of the distance data information is maximally preserved. The time complexity of our accelerated uniform spherical sampling method is then improved to be $O(n)$.

Uniform Impingement Detection and Surface-to-Surface Distance Measuring Method

During our ball-and-socket joint simulation, we process the impingement detection scan twice: one model's vertices against the other model's LUT. As an example, we do a scan which tests the ball's vertices against the socket's LUT. We can compute the spherical coordinates (ϕ and θ) of a target vertex v on the ball, compute the interpolated distance d from the neighboring *helix orientation points* on the spherical helix, and compute the distance d_v from the ball vertex to the same system origin. The collision detection can be instantly done by comparing the difference between d and d_v to find if the vertex is inside or outside of the socket surface.

We need to locate the *helix orientation points* neighboring to the target vertex v to get the sampled distance data from the LUT (Fig. 6a). There are two spherical helix turns that neighbor to v and four *helix orientation points* on these turns that have the closest orientations to this vertex. Two neighboring *helix orientation points* are located in the turn above and two are in the turn below. We can find the neighboring *helix orientation points* through the following three steps:

a) Given the spherical coordinate ϕ_v of v, we can first easily locate the two neighboring turns based on Eq. 2.
b) Then given spherical coordinate θ_v of v, we can find two intersection points where the line $\theta = \theta_v$ intersects with the two turns.
c) Finally, in each turn two *helix orientation points* closest to the intersection point can be located.

For example, in Fig. 6b P_a is v's intersection point with the turn above while P_b is v's intersection point with the turn below. P_0 and P_1 are the *helix orientation points* neighboring to P_a while P_2 and P_3 are the *helix orientation points* neighboring to P_b. Therefore, P_0, P_1, P_2, and P_3 are the *helix orientation points* neighboring to v. There is one exceptional case when the vertex is located inside the region of the first turn or the last turn. We just need to find three neighbors.

Steps a) and b) can be easily done. The key step is step c). Efficiently finding the two *helix orientation points* neighboring to the intersection point can be done by comparing the spherical helix points' *Traveled Distance T* (Fig. 6c):

$$SHPointTraveledDistance\ T = \frac{(4\times \mathrm{k}+1)\times(-1\times\cos(\phi)+1)}{2} \tag{3}$$

where ϕ is the spherical coordinate of the point and *k* is the number of turns in the helix

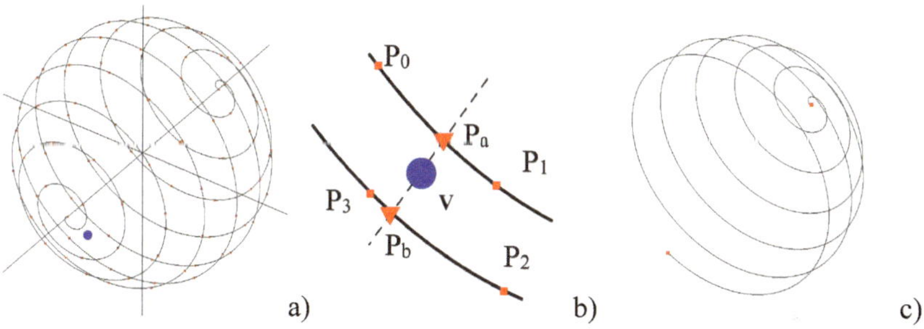

Fig. 6. Locating the sample points close to the orientation of vertex *v* (in *blue*): (**a**) global view of the spherical helix curve (*k*=10); (**b**) local view; and (**c**) one example of traveled distance. The black curve indicates P_b's traveled path from the start point of the spherical helix curve.

Equation 3 is gained from data fitting based on Eq. 2 for the connection between a *k*-turn spherical helix point's segment length and the point's coordinate ϕ. Each spherical helix point, including the *helix orientation points* specified by Eq. 2, has its *traveled distance T*. The *traveled distances* of the spherical helix points are used to examine the points' neighbor relationships on the spherical helix curve.

With the intersection point's ϕ coordinate, the intersection point's *traveled distance T* can be instantly computed from Eq. 3. Since all *helix orientation points* are equally distributed along the curve with a distance Δx, the LUT index numbers of the two neighboring points to the intersection point are ($floor(T / \Delta x)$) and ($floor(T / \Delta x)+1$).

After all four neighboring points are located, the interpolated distance data *d* in *v*'s orientation can then be computed based on these neighboring *helix orientation points*. The impingement detection can be done by examining the difference of sampled distance *d* and the vertex's distance d_v. This distance difference can be saved as the surface-to-surface distance in *v*'s orientation and can be further referred to as the approximate strain or the impingement level in the medical simulation.

Algorithm 1: impingement detection algorithm of USID system (per iteration)

For i=0; i<number of model B's vertices; i++
  Convert vertex B_i in model A's local spherical coordinate system;
  Compute distance d_i from B_i to the system origin;
  Compute spherical coordinates [ϕ_i, θ_i];
  Locate the neighboring turns of the helix curve based on ϕ_i;
  Locate 2 neighboring spherical helix points in each turn based on θ_i;
  Compute interpolated distance d in [ϕ_i, θ_i] based on A's LUT;
  if $d <= d_i$
    Impingement detected on that vertex;
    $\Delta d = d - d_i$;
    Save Δd as the surface-to-surface distance in the orientation [ϕ_i, θ_i];
  else
    There is no impingement;

The complexity of uniform impingement detection is $O(n)$ where n is the number of the model B's vertices to be tested against model A's LUT.

Results

All tests were conducted using an AMD Opteron Processor 252 at 2.6 GHz with 2 GB of memory. A set of real models were segmented from CT data to validate the practicability and reliability of the system (Fig. 7). The USID system was applied on several resolutions of the models from 300-triangle level to 12k-triangle level to develop and prove the design. Notice that in practice two LUTs are created for the socket and the ball. USID are applied twice to avoid inter-penetration artifacts.

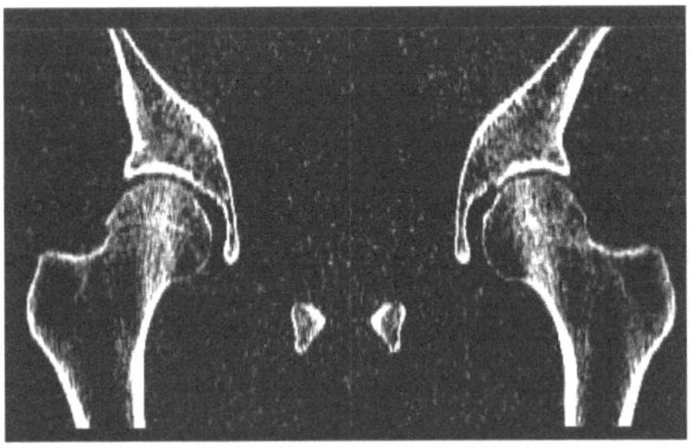

Fig. 7. The patient's pelvis region is scanned into a set of 147 CT slices (1.25 mm thickness, resolution of 512 × 512), anterior view.

Before the tests, we need to configure the sampling precisions for USID and match USID system with RSID system to produce comparable results. For RSID, we choose three sampling precision levels from 1° to 0.05° as shown in Table 1; for USID, the same sampling precision can be configured by a proper number of the turns *k*. For each precision, *k* can be determined by matching the numbers of the sample points at the equator generated by RSID and USID. The equator has lowest point density among all ϕ slices (or latitudes) in non-uniform sampling, which means the point density at the equator is the bottom line to determine the sampling accuracy or the sampling error. In Table 1, the non-uniform sampling precisions and the corresponding *k* are matched. Table 2 lists the comparisons on the memory cost of our non-uniform RSID system and our USID system.

Table 1. Three sampling precisions in RSID and their corresponding *k* in USID.

Precision in RSID	Turns of helix (*k*) in USID
precision=1° (360 points at equator)	*k*=229 (360 points at equator)
precision=0.1° (3,600 points at equator)	*k*=2,292 (3,601 points at equator)
Precision=0.05° (7,200 points at equator)	*k*=4,584 (7,201 points at equator)

Table 2. The memory requirement of non-uniform RSID system and USID system under three sampling precision settings.

Precision	Size of 1D LUT in USID system	Memory requirement of RSID (MB)	Memory requirement of USID (MB)
1°	$w = k^2$=52,441	0.497	0.40
0.1°	$w = k^2$=5,253,264	49.5	40.07
0.05°	$w = k^2$=21,013,056	197.8	160.32

The comparisons show USID can reduce about 20% sample points or 20% memory cost. In Fig. 8, the global density of the sample points generated by RSID and USID is compared. We observed that in the area around the pole the sample point density has a clear reduction as desired.

Our non-uniform sampling method has an efficiency of $O(n)$ and makes the preprocess step much shorter. The uniform version of sampling takes more computations for a fair memory cost reduction but the sampling speed is still fast when it is accelerated by the sampling algorithm designed for the RSID system (Table 3).

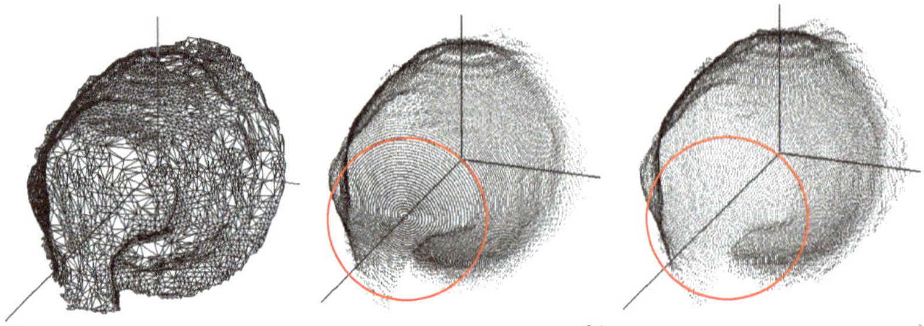

Fig. 8. The comparison of sample point density generated by our two sampling methods at same sampling precision=1°: (**a**) the original socket model; (**b**) its sample points (65,160 points) generated by our non-uniform sampling; (**c**) its sample points (52,441 points) generated by our uniform sampling.

Table 3. Benchmark of three sampling methods on 3k triangle-level models only.

Precision	Pair-wise testing, $O(n^2)$ (ms)	Sampling method in RSID, $O(n)$ (ms)	Sampling method in USID (accelerated) $O(n)$ (ms)
1°	176,328	15	31
0.5°	700,844	16	109
0.1°	17,722,156	62	2,781
0.05°	N/A	187	11,172

The impingement detection algorithm in USID requires higher computation cost than RSID because it takes more steps analyzing the spherical helix curve to locate the neighboring sample points on the curve. The benchmarks in Table 4 show the results; however, the impingement detection algorithm in the USID system is still an $O(n)$ algorithm and the efficiency, as shown in Table 4, is sufficient for real-time simulations. The impingement detection method in USID is also compared to RAPID [7] in Table 4.

Table 4. Benchmark of four impingement (collision) detection methods on three different triangle levels.

Triangle level (per model)	Pair-wise testing $O(n^2)$ (ms)	RAPID [7] (ms)	RSID $O(n)$ (ms)	USID $O(n)$ (ms)
300-triangle	30.7	1.41	0.06	0.07
3k-triangle	2,831	1.43	0.48	1.8
12k-triangle	N/A	1.45	5.7	12

The triangle level in Table 4 shows the number of triangles on each of the femoral head and the acetabulum. The models were set up based on their original relative coordinates from the CT data. Notice that the comparisons in Table 4 are just speed-only tests to prove the efficiency level that our system is suitable for real-time simulations. Other important features are not concerned in the test. RAPID is aiming on the speed with its OBB-Tree efficiency and only reports the existence of collisions; however, it is not able to provide distance information between two surfaces while the benchmarks of our methods, RSID and USID, have already included the computation cost on the surface distance measurement. Surface-to-surface distance measurement is considered as one major goal of our simulation, which is referred to estimate the approximate strain distribution and the impingement level. This feature can provide more important overall examination than just checking the existence of the collision by marking a few collided points. One example of impingement detection results with distance measurement is illustrated in Fig. 9.

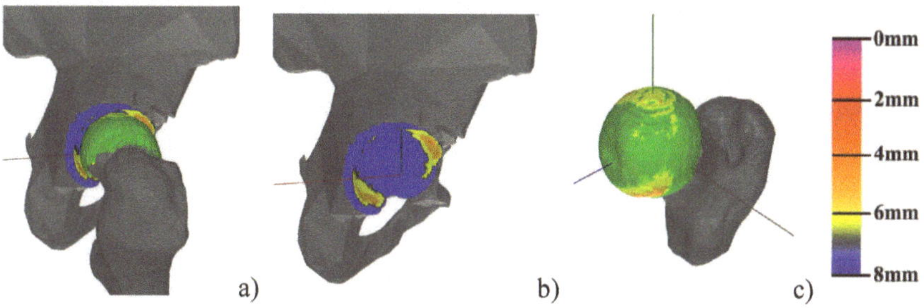

Fig. 9. USID simulation surface-to-surface distance measurement applied on a real hip joint model segmented from CT data (1.25 mm slice thickness, resolution of 512x512).

Conclusion

In this chapter, we presented a novel efficient and accurate impingement detection system with rapid uniform sampling and useful surface-to-surface distance measurement feature for ball-and-socket joint. Comparing with most of the current joint impingement detection methods, our system has the advantages on the efficiency, the accuracy, and the ability to provide more detailed results. The system has memory-efficient feature and this advantage is more significant for high-precision models or for multiple models in one scene. The rapid and accurate impingement detection with surface-to-surface distance measurement can provide more realistic detailed information to estimate the overall impingement on the ball-and-socket joint, which is particularly useful for computer-aided surgical plan.

Acknowledgments This project is funded by NSERC Discovery grant. We appreciate Andrew Speirs (Division of Orthopaedic Surgery, Ottawa Hospital-General Campus) for the advices on CT data segmentation tool, Matt Kennedy (Biomedical Engineering, University of Ottawa) for the range of motion data, and Anna Conway (Division of Orthopaedic Surgery, Ottawa Hospital-General Campus) for coordinating CT data collection.

References

1. Ganz R, Parvizi J, Beck M, Leunig M, Notzli H, Siebenrock KA (2003) Femoroacetabular impingement: A cause for osteoarthritis of the hip. Clinical Orthopaedics and Related Research 417:112–120
2. Wisniewski SJ, Grogg B (2006) Femoroacetabular impingement: An overlooked cause of hip pain. American Journal of Physical Medicine and Rehabilitation 85:546–549
3. Teschner M, Kimmerle S, Heidelberger B, Zachmann G, Raghupathi L, Fuhrmann A, Cani MP, Faure F, Magnenat-Thalmann N, Strasser W, Volino P (2005) Collision detection for deformable objects. Computer Graphics Forum 24(1):61–81
4. Kockara S, Halic T, Iqbal K, Bayrak C, Rowe R (2007) Collision detection: A survey. Systems, Man and Cybernetics, ISIC. IEEE:4046–4051
5. O'Sullivan C, Dingliana J (1999) Real-time collision detection and response using sphere-trees. In: Proceedings of the Spring Conference in Computer Graphics, Bratislava. pp 83–92
6. Dinesh M, Jonathan DC, Stefan G (1996) Collision detection: Algorithms and applications. Algorithms for Robot Motion and Manipulation 129–142
7. Gottschalk S, Lin M, Manocha D (1996) Obb-tree: A hierarchical structure for rapid interference detection. In: Computer Graphics (SIGGRAPH '96), pp. 171–180
8. Fuhrmann A, Sobottka G, Groß C (2003) Distance fields for rapid collision detection in physically based modeling. In: International Conference on Computer Graphics and Vision, Keldysh Inst. of Applied Mathematics, pp. 58–65
9. Sud A, Otaduy MA, Manocha D (2004) DiFi: Fast 3D distance field computation using graphics hardware. Computer Graphics Forum (Proc. Eurographics) 23(3):557–566
10. Sigg C, Peikert R, Gross M (2003) Signed distance transform using graphics hardware. In: IEEE Visualization 2003, IEEE Computer Society Press, pp. 83–90
11. Sarni S, Maciel A, Boulic R, Thalmann D (2004) Evaluation and visualization of stress and strain on soft biological tissues in Contact. In: International Conference on Shape Modeling and Applications, pp. 255–262
12. Yoshida H, Faust A, Wilckens J, Kitagawa M, Fetto J, Chao E (2006) Three-dimensional dynamic hip contact area and pressure distribution during activities of daily living. Journal of Biomechanics 39(11):1996–2004
13. Hu Q, Langlotz U, Lawrence J, Langlotz F, Nolte LP (2001) A fast impingement detection algorithm for computer-aided orthopedic surgery. Journal of Computer Aided Surgery 6(2):104–110
14. Kubiak-Langer M, Tannast M, Murphy SB, Siebenrock KA, Langlotz F (2007) Range of motion in anterior femoroacetabular impingement. Clinical Orthopedics and Related Research 458:117–124
15. Maciel A, Boulic R, Thalmann D (2007) Efficient collision detection within deforming spherical sliding contact. IEEE Transactions on Visualization and Computer Graphics 13(3): 518–529
16. Praun E, Hoppe H (2003) Spherical parametrization and remeshing. In: ACM SIGGRAPH 2003, pp. 340–349
17. Nishio H, Altaf-Ul-Amin Md, Kurokawa K, Kanaya S (2006) Spherical SOM and arrangement of neurons using helix on sphere. IPSJ Transactions on Mathematical Modeling and Its Applications 47:56–60
18. Cai D, Lee WS, Joslin C (2008) Rapid ball-and-socket joint collision detection. In: IEEE International Workshop on Medical Measurements and Applications, pp. 25–28

X-ray-Based Craniofacial Visualization and Surgery Simulation

Junjun Pan[1], **Jian J Zhang**[2], **Yanning Zhang**[3], **and Hong Zhou**[4]

Abstract X-rays penetrate both soft and hard tissues, which record only the accumulated density value of tissues. It is therefore impossible to represent the geometry of human tissues accurately with a small number of X-rays. CT can acquire the geometry of human anatomy accurately. It, however, subjects the patient to a high dose of radiation which in many cases is undesirable and unhealthy. This chapter presents a novel craniofacial visualization technique with the developments from both computer graphics and computer vision. It is a low-radiation, low-cost alternative to CT-based system for the reconstruction of 3D cranium using only three X-rays. We paste lead markers on the subject's face which allow a 3D face model to be constructed using correlated vision. Then the surface of the cranium is obtained by subtracting the soft tissue depth from the face surface. Because of the penetrating nature of X-rays, existing computer vision techniques are not effective in matching the corresponding points for X-rays. We present a new matching algorithm to solve this problem by evolutionary programming. We also designed a supervised learning method to estimate the soft tissue stiffness parameters.

Introduction

3D cranium reconstruction has received a lot of attention from both medical and computer graphics research communities and is an effective approach for the visualization of the complex internal cranial structure. Existing methods acquire the raw image data by CT scanning, with which the volumetric geometry is reconstructed to produce a 3D computer graphics (CG) model of the head. They have

[1] Bournemouth University, National Center for Computer Animation, Poole, UK
pjunjun@bournemouth.ac.uk

[2] Bournemouth University, National Center for Computer Animation, Poole, UK
jzhang@bournemouth.ac.uk

[3] Northwestern Polytechnical University, School of Computer Science, Xi an, China
ynzhang@nwpu.edu.cn

[4] Stomatology Hospital of Xi an Jiao tong University, Xi an, China
zhou57@pub.xaonline.com

N. Magnenat-Thalmann et al. (eds.), *Recent Advances in the 3D Physiological Human*,
DOI 10.1007/978-1-84882-565-9_13, © Springer-Verlag London Limited 2009

been widely used in various medical visualization applications, including surgical planning, diagnosis and medical training.

Although it is justifiable to scan a subject by CT for serious diseases, such as tumour or accident-caused cranial trauma, the high dose of X-ray of CT scan can in many cases deter medical practitioners from taking advantage of the latest technology in computer graphics and visualization research. Taking orthodontic for young children as an example, CT scans are not very often used. One of the reasons is the X-ray dosage the subject receives. Statistics [1] suggests youngsters make up about 80% of the patients in orthodontics and maxillofacial surgeries. And the X-ray dose from one CT head scan is approximately 20 times of the dose in one X-ray photographing. Large dose of X-ray radiation could have a serious adverse effect to the juvenile patient population. In addition, in comparison with ordinary X-ray photography (cephalometry), a CT scan is also more costly, which could be an issue for developing countries and underprivileged regions. In practice, ordinary X-ray photographing remains by far the most accessible and most commonly used medical imaging tool in craniofacial clinics. The technological advancement of 3D computer graphics has not been fully exploited in common medical practices. This is especially the case for orthodontics where the correction of the bony structure is usually followed and checked for years while the subject grows in age, as one would not like to subject the young patients with regular high dosage of X-ray radiation.

In this chapter, we present a novel craniofacial visualization technique. Although at this stage, it remains a distance from being a clinically viable alternative, we hope this idea would merit further attention and development due to its unique advantages over the traditional X-ray-based approach. We make three contributions in the chapter. The first is a new craniofacial reconstruction technique combining the knowledge of computer graphics and computer vision. Instead of relying on a CT scan to acquire the internal structure of the head, we use only three ordinary X-ray photographs (cephalograms) together with an ultrasound scan. Compared with a full-blown CT scanning system, our technique is able to produce a CG skull and the facial skin surface incurring only a fraction of the X-ray exposure to the subject. The second contribution is a new corresponding point matching algorithm that is able to identify corresponding feature points on different X-rays. In order to produce reasonably accurate simulation results, our third contribution is an easy-to-implement parameter estimation technique used with our finite element model (FEM) for the simulation of soft tissue deformations.

Human skulls and faces are of a free form and complex shape. A small number of ordinary X-rays do not provide sufficient information for their geometry to be derived, owing to the penetration nature of X-rays. In order to overcome this problem, we stick a number of lead markers on the subject's face and take three X-ray photographs. The images of these lead markers on the three X-ray photos allow the face surface to be reconstructed once the correlated vision of the markers is established. The X-ray images we take display only the profiles of the skull. As they only record the accumulated density value of the tissues, not the geometry, three profiles curves of skull are far from enough for a detailed model to be rebuilt.

Therefore, acquiring the facial skin surface is essential, not only for the evaluation of the appearance of the subject but also for the reconstruction of the skull surface. With the subject's face model, we then derive the skull surface by subtracting the soft tissue thickness, which can be measured by M-ultrasound [2]. The profile curves of the skull, obtained by cephalograms, provide ground truth for the reconstruction, which is very useful to calibrate our ultrasound measurement.

The simulation of soft tissue deformations has also attracted a great deal of research efforts and there exist a number of techniques. Two factors are important for the simulation accuracy, which are the mathematical model and the material parameters pertaining to the mathematical model. In this chapter, we present a simple, but practical method for the estimation of the relevant simulation parameters, which are essential for enhancing the simulation accuracy. The estimated parameters are then used in our double-layered finite element model for the simulation of the soft tissue deformations.

This chapter is structured as follows. After a brief introduction of the related work in the second section, in the third section we describe our X-ray-based craniofacial reconstruction and visualization technique including our new point matching algorithm. The simulation model and a stiffness parameter estimation technique are discussed in the fourth section. The fifth section explains the experiments to study the accuracy and effectiveness of the approaches. We discuss the limitations of our system and future improvements in the sixth section.

Related Work

There are many craniofacial measurement and analysis methods, each has advantages and disadvantages and is appropriate for certain applications. In modern medical imaging, tomography, such as CT or MRI, represents the dominating technology in the reconstruction of the cranium with high accuracy [2, 3]. Recently, research on 3D craniofacial visualization and surgical simulation using both CT and laser scanning techniques has gained a popularity, which achieves high accuracy in the measurement of both soft tissue and bony structures [4–8]. Reconstructing a head model using CT scans is relatively trivial, which can be undertaken with various volumetric graphics packages [2]. Because of the complexity of the human head structure and tissue behaviours, accurate simulation of tissue deformations proves very challenging. Many techniques have been explored by the research community with a varying degree of success. The typical deformation models include the finite element (FE) simulation [9, 10], mass–spring systems (MSS) [11] and the Chain-Mail technique [12]. Among them, the FE models are likely to produce the best simulation accuracy, albeit its dependency on parameters setting. The advantage of the Chain-Mail algorithm is its ability to achieve a very high computational efficiency.

Although CT imaging produces high accuracy, some crucial factors, such as radiation and cost of the equipment, are of a concern to its applications. Some

Japanese and Chinese medical institutes have realized this issue and made an effort to develop 3D cranium analysis methods from multiple orthodontic X-rays [13]. It works by manually selecting a set of feature points on the X-ray images and calculating the 3D coordinates of them resulting in a craniofacial measurement. However, this system can only measure the given feature points and is unable to build a 3D cranium model. Laser scanning and structured lights are other effective non-invasive methods in the reconstruction of 3D human faces. However, they cannot measure the depth of soft tissues. If combined with X-rays, both types of the images have to be correctly registered, which is another tough problem in medical image processing.

Reconstruction of Face and Skull

X-ray imaging belongs to the penetrating projection imaging model. Each pixel on the image plane integrates the density of all the voxels through the line of sight. This means that an X-ray image does not offer any 3D surface or geometry information. Because of this, one is not able to derive a full geometric model from only a small number of X-ray photos, unless a full CT scan is used.

Instead of trying to model a skull from X-ray photos directly, we endeavour to identify the skull surface starting from the face geometry. We use three X-ray photos for the purpose of both reconstructing a face surface and providing a datum for skull modeling. X-rays penetrate both soft and hard tissues, but not the metal lead. This makes the lead marker points visible on the X-rays, which are placed on the face of the subject. We calculate the 3D position of lead markers using correlated vision. The details of our point matching technique are given below.

Our technique consists of five steps: (1) enhancing images for easy detection of the pixels associated with markers; (2) identifying a feature region on the X-ray images; (3) matching corresponding feature regions on different X-ray images; (4) reconstructing a face model based on the computed 3D coordinates of all markers; and (5) deriving the skull surface by subtracting the soft tissue thickness that is measured with M-ultrasound scanning.

System Structure

The structure of our craniofacial visualization and simulation system is illustrated in Fig. 1.

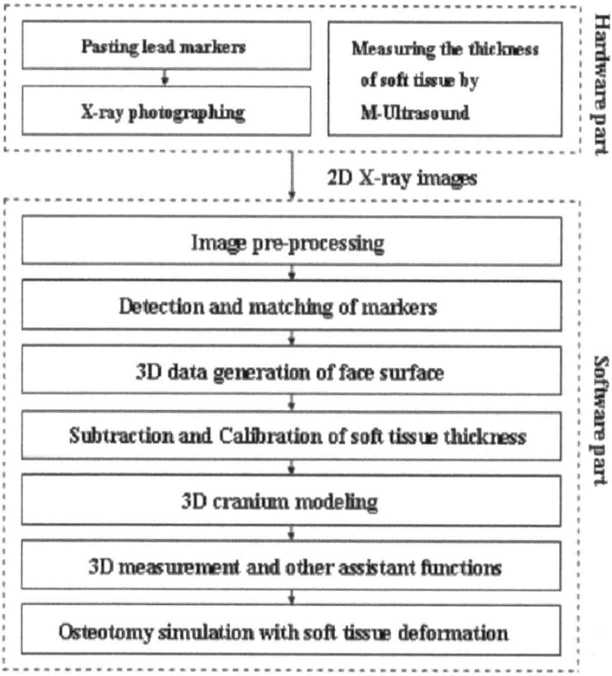

Fig. 1. Structure of X-ray-based craniofacial visualization and surgery simulation system.

To produce the face model for a subject, we first paste a number of lead markers on the face of the subject, which are fixed on an adhesive tape beforehand. In our experiment, the lead markers are placed in a 30×15 matrix for half a face. We then fix the subject's head in the cephalometer in order to satisfy the requirements of the correlated vision. Experiments show that the optimal projection orientations are 0° (frontal), 60° and 90° (lateral) because of the least occlusion. We take three X-ray photos from these orientations. The X-ray images of 60° and 90° are used to reconstruct the 3D face model and those of 0° and 90° are used to measure the thickness of soft tissue and calibrate the M-ultrasound scanning data. To derive the accurate thickness of the soft tissue, the technician marks the approximate location of key markers on the skin by crayon after the adhesive tape is removed. The soft tissue thickness is measured using M-ultrasound, probing on the marked points on the face. The whole process usually takes 5–6 minutes. It is inevitable that errors are likely to be introduced from the manual operations. However, the two calibrating X-rays taken at 0° and 90° provide accurate profile curves of the skull. This step allows the M-ultrasound measurements to be normalized and errors minimized. Figure 2 shows a sketch of the X-ray photographing process.

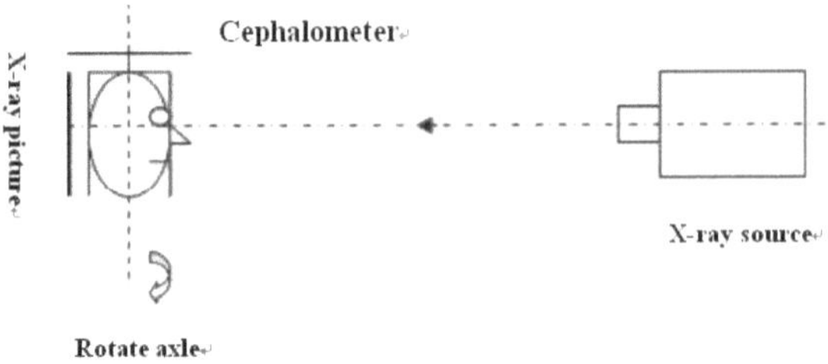

Fig. 2. X-ray photographing with cephalometer.

Surface Imaging and Feature Point Recognition

Two original X-ray images are shown in Fig. 3.

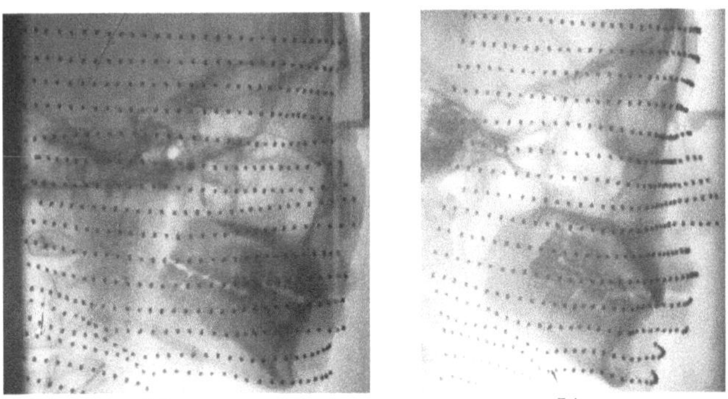

(a) (b)

Fig. 3. Original X-ray images pair: (**a**) 60° projection; (**b**) 90° projection.

After histogram equalization, a following high-pass filter mask is designed to enhance the contrast of the isolated regions (the marker regions).

$$\begin{bmatrix} -5 & -3 & -1 & 1 & -1 & -3 & -5 \\ -3 & 0 & 1 & 2 & 1 & 0 & -3 \\ -1 & 1 & 2 & 4 & 2 & 1 & -1 \\ 1 & 2 & 4 & 8 & 4 & 2 & 1 \\ -1 & 1 & 2 & 4 & 2 & 1 & -1 \\ -3 & 0 & 1 & 2 & 1 & 0 & -3 \\ -5 & -3 & -1 & 1 & -1 & -3 & -5 \end{bmatrix}$$

Figure 4 gives the processed result. With the enhanced image, the feature points can be easily segmented by thresholding grey levels. To identify the image of a lead feature point on an X-ray photo, we designed a simple image recognition technique. Although numerous pattern recognition techniques exist, ours is intuitive and straightforward to use.

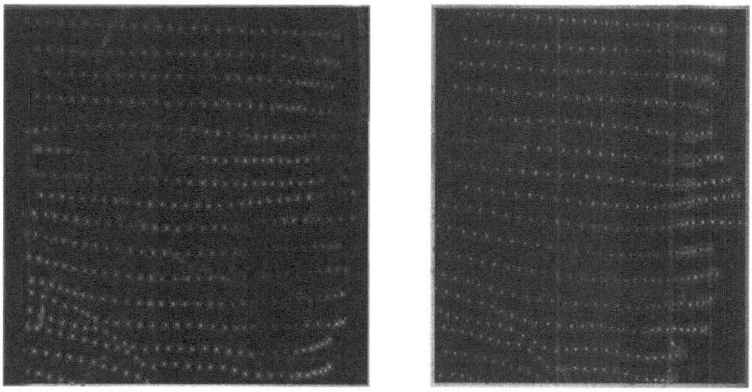

(a) (b)

Fig. 4. Image pair after feature points enhanced: (**a**) 60° projection; (**b**) 90° projection.

Our feature-enhancing treatment above (Fig. 4) has led to the arrival of a binary image, where each pixel has either a 0 or 1 value. The feature region detection algorithm simply visits all pixels on a row-wise basis. A feature region on the image is identified if the value changes between adjacent pixels. In order to avoid the disturbance of image noise, a detected region is confirmed only if it spans at least a given number of pixels. We assume a region to be at least 6 pixels across in our experiments.

Corresponding Points Matching

Having detected the feature points, the next task is to match the corresponding points on two different X-ray photos to compute their 3D coordinates. Basically, a point pair should be searched along the same epipole line. This is a typical computer vision problem and there are many matching algorithms available [14–18], which can be mainly divided into two types: grey level based and geometric feature based. However, due to the penetrating projection nature of X-rays, neither of the two types is effective in this problem [17].

To tackle this issue, we present a new point matching algorithm, which is based on point position similarity, as described as follows:

In a correlated image pair, suppose there are n feature points in each image. Scan this image pair in the same order. During scanning, when a feature point is detected, its 2D coordinates are stored in an array. Scanning both images results in two arrays to be created containing the 2D coordinates of the feature points. Suppose the array for the first image is

$$L\{l_1(x,y), l_2(x,y), l_3(x,y), \ldots, l_{n-1}(x,y), l_n(x,y)\} \tag{1}$$

and that for the second image is

$$R\{r_1(x,y), r_2(x,y), r_3(x,y), \ldots, r_{n-1}(x,y), r_n(x,y)\} \quad (2)$$

For each feature point in one image, there is only one corresponding point in the second image regardless of the scanning order. This is a permutation problem and there are $n!$ different answers, among which only one is correct. Due to occlusion, markers may be missing from the images occasionally. But our experience did not find this poses a problem. The missing points can be easily added by the operator manually to make the total feature point number the same for both left and right images.

When the difference of the projection angles is small for both correlated images and the object, on which the feature points are placed, is largely convex, adjacent feature points have the same relative positions (or orders) on both images, i.e. they satisfy the ordering constraints in local region. (The geometry meaning of the ordering constraints is if $l_i(x,y)$ is a point in the first image and $l_j(x,y)$ is another point in the adjacent area of $l_i(x,y)$, their corresponding points in the second image are $r_i(x,y)$ and $r_j(x,y)$. Suppose $x_{l_i} \le x_{l_j}$, $y_{l_i} \le y_{l_j}$ exists, then there will be $x_{r_i} \le x_{r_j}$, $y_{r_i} \le y_{r_j}$.) Place both correlated images in the same coordinate system, and let the barycentre of feature points in both images coincide. Our proposition is that the summation of the Euclidean distance between pairs of corresponding points becomes minimized if the correct matching is found. We call this algorithm the shortest distance matching based on position similarity. Thus we have

$$\min \sum_{i,j=1}^{n} d_{ij}\delta_{ij}$$

$$\textbf{s. t.} \quad \sum_{j=1}^{n} \delta_{ij} = 1,\ i = 1,2,3,\ldots,n,$$

$$\sum_{i=1}^{n} \delta_{ij} = 1,\ j = 1,2,3,\ldots,n, \quad (3)$$

$$\delta_{ij} \in \{0,1\}, i,j = 1,2,3,\ldots,n,$$

$$d_{ij} = \sqrt{(x_{li} - x_{rj})^2 + (y_{li} - y_{rj})^2}, i,j = 1,2,3,\ldots,n,$$

$\delta_{ij} = 1$ suggests the number i point in array $\boldsymbol{L}$ matches the number j point in array $\boldsymbol{R}$. d_{ij} stands for the Euclidean distance between point i in array $\boldsymbol{L}$ and point j in array $\boldsymbol{R}$. The objective function minimizes the summation of the distances between two matched points. The two constraints represent that each point in one image should match only one point in the other image.

We solve this optimization problem by evolutionary programming. Using the reciprocal of $\min \sum_{i,j=1}^{n} d_{ij}\delta_{ij}$ to measure the fitness, the algorithm can be processed in the following steps: making $\boldsymbol{L}$ match $\boldsymbol{R}$ in order. First we fix the sequence of the feature points in array $\boldsymbol{L}$, change the sequence of two feature points in array $\boldsymbol{R}$ when there is a mutation operation in this position. Then calculate the fitness of each individual in new matching result, and eliminate the individuals whose fitness is small. The algorithm ends when a required tolerance (e.g. the difference of fitness) is reached leading to the identification of points matching. In the next section, we will compute the 3D coordinates of these marker points with stereoscopic method from which a 3D face model can be generated.

Coordinates Computation and 3D Face Generation

Two rigid-body transformations should be used in this step:

$$\begin{pmatrix} x \\ y \\ z \end{pmatrix} = \begin{pmatrix} x_l \\ y_l \\ z_l \end{pmatrix} + t_l \mathbf{R}_l \begin{pmatrix} x'_{l,i} \\ y'_{l,i} \\ F_l \end{pmatrix}, \quad \begin{pmatrix} x \\ y \\ z \end{pmatrix} = \begin{pmatrix} x_r \\ y_r \\ z_r \end{pmatrix} + t_r \mathbf{R}_r \begin{pmatrix} x'_{r,i} \\ y'_{r,i} \\ F_r \end{pmatrix} \tag{4}$$

$\mathbf{R}_l$ and $(x_l, y_l, z_l)^T$ are the rotation matrix and translation vector for the left image (60° projection). Similarly, $\mathbf{R}_r$ and $(x_r, y_r, z_r)^T$ are the rotation matrix and translation vector for the right image (90° projection). F_l (F_r) stands for the focal length of the left (right) image, t_l and t_r are two unknowns to be determined, which can be obtained by minimizing d in (5).

$$d^2 = \left[\begin{pmatrix} x_l \\ y_l \\ z_l \end{pmatrix} + t_l \mathbf{R}_l \begin{pmatrix} x'_{l,i} \\ y'_{l,i} \\ z'_{l,i} \end{pmatrix} - \begin{pmatrix} x_r \\ y_r \\ z_r \end{pmatrix} - t_r \mathbf{R}_r \begin{pmatrix} x'_{r,i} \\ y'_{r,i} \\ z'_{r,i} \end{pmatrix} \right]^2 = [\mathbf{b} + t_l \mathbf{r}_l - t_r \mathbf{r}_r]^2 \tag{5}$$

Where $\mathbf{b}$ stands for the basis of the world coordinate system. $\mathbf{r}_l$ ($\mathbf{r}_r$) denotes the left (right) ray of sight rotated into the world coordinate system. Once t_l and t_r have arrived, the 3D coordinates of the feature point can be calculated using (4). The surface model is then trivially derived from the marker points. Figure 5a shows the 3D face model of the subject in the experiment in the Fig. 3. Notice that the reconstructed face model depends on where the markers are placed. As markers are not distributed, the eyes, eyebrows and hair cannot be modeled properly. Fortunately, they are not necessary for the medical measurement we aim for.

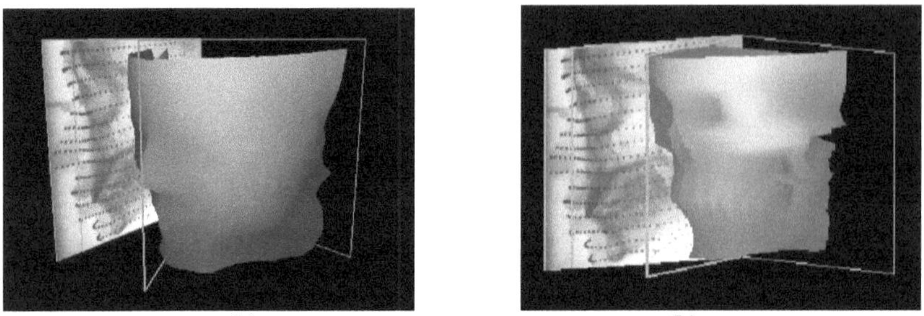

(a) (b)

Fig. 5. 3D face and cranium model of a subject in experiment.

3D Cranium Generation

From the obtained face model and using the thickness data by M-ultrasound scanning, we project the feature points inwards to derive the geometry of the skull. Generally, the resolution of X-rays is much higher than ultrasonograph. Therefore, to make the measurement more accurate, the M-ultrasonograph data need to be normalized by the soft tissue depth in lateral and frontal X-rays (90° projection and 0° projection) first. This calibration process can be computed by the following formula:

$$D_{i,j} = d_{i,j}\sqrt{\sin^2\theta(\frac{D_{ni}}{d_{ni}})^2 + \cos^2\theta(\frac{D_{0i}}{d_{0i}})^2} \tag{6}$$

where $d_{i,j}$ is the original depth measured by M-ultrasonograph in the marker position of ith row, jth column. θ is the orientation angle in this position. d_{0i}, d_{ni}, respectively, represent the original ultrasound measured depth in the position of the first and last markers at the ith row. D_{0i}, D_{ni} represent their corresponding depth in frontal and lateral X-rays. $D_{i,j}$ is the normalized depth. We assume the soft tissue thickness measurements were taken in the normal direction to the face skin surface. Figure 5b shows the 3D cranium surface model of the subject which is calculated from the face model shown in Fig. 5a.

Surgery Simulation and Soft Tissue Deformation

Once the skull and face models are generated, we can then implement a surgical simulation system. Soft tissue deformation is a complex phenomenon and its accurate simulation is of a practical significance. Many simulation models exist, which endeavour to approximate the soft tissue deformations as accurately as possible for virtual surgeries. Broadly speaking, there are two main physically based approaches, one is based on a finite element model (e.g. [9,10,19,20]), and the other one based on mass spring systems [11]. Both FEM and MSS have had a varying degree of success in the simulation of soft tissue deformations. Since an accurate simulation proves extremely challenging due to the complex behaviour of living human tissues, all current techniques only attempt to approximate the behaviours.

Our FEM simulation method is not technically novel. Our contribution in this section is concerned with the formulation of the simulation parameters accurately. Here we briefly present our simulation model and discuss our supervised learning technique for the acquisition of accurate simulation parameters.

Before describing our soft tissue deformation and surgery simulation technique, let us first introduce some relevant medical background. Our study focuses on orthodontics, although the developed technique is applicable to other craniofacial visualization and simulation purposes. In orthodontics there are mainly two typical clinic abnormalities. One is known as the LeFort I, a kind of abnormality of the maxilla, which needs LeFort I pattern osteotomy. The other is the abnormality of the jawbone, which needs mandible vertical osteotomy. In Fig. 6a, c, the red broken line illustrates the position and direction of bone cutting. Figure 6b, d shows the operation simulation of these two typical osteotomies in our system.

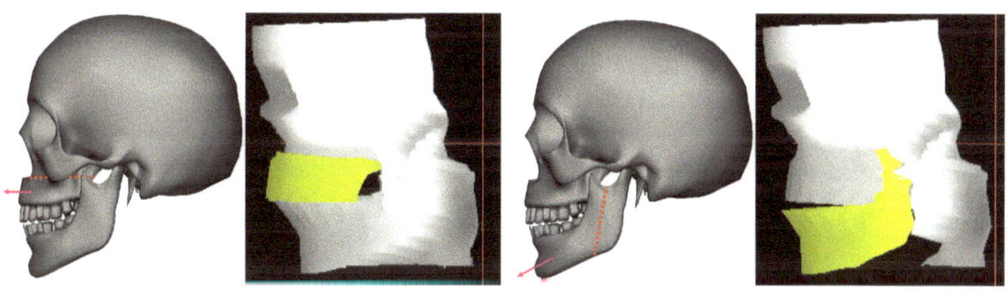

(a) (b) (c) (d)

Fig. 6. Two typical osteotomies and corresponding operation simulations: (**a**) LeFort I osteotomy; (**b**) surgery simulation of LeFort I; (**c**) mandible vertical osteotomy; (**d**) surgery simulation of mandible vertical osteotomy

To model the deformation behaviour of the soft tissues subject to surgical manipulations, we develop a double-layered non-linear, globally C^1 continuous finite element model for the face and cranium. The soft tissue forms the outer layer and the bone tissue is represented as the inner layer. It uses quadrangle polynomial shape functions similar to the modeling paradigm mentioned in [9]. Two material parameters, the modulus of elasticity (E) and Poisson's ratio (υ), are important for the simulation accuracy. Material properties vary from person to person and vary with the conditions of the person concerned, such as age, gender, ethnic origin and the environment they live in. This variation makes an accurate simulation model extremely difficult to come by. However, this issue appears to have been somewhat neglected in the literature.

In this research we hypothesize that the stiffness parameters are largely invariable in the local part of the facial surface for the same subject. Clinical practice suggests that facial deformities, such as the deformities for orthodontics, often require multiple corrective surgical interventions. In the following, we present an easy-to-implement supervised machine-learning method to calculate the stiffness parameters based on some measurements obtained from the first operation. Although our hypothesis seems restrictive, it is very helpful for the planning and prediction of subsequent surgical operations, as multiple operations are often necessary in orthodontics. Our simulation technique based on the above-presented cranial reconstruction technique gives the surgeon confidence in predicting the resultant outcomes with accuracy.

The training model adopts an iterative strategy, as given in (7) below. At the beginning, the parameters are initialized with guess values. If available, one can use the parameters from other patients with similar conditions. After the computation of the soft tissue deformations using the above-proposed finite element model, the results are compared with the measured reference data (training data) taken from the operation. The errors are used to adjust the estimated parameters and this training process goes on until a predefined tolerance is satisfied. The training formula is given by

$$\varepsilon_{i+1} = \varepsilon_i + (T - t_i)k \tag{7}$$

where ε_{i+1} is one of the material parameters in step $i+1$ and ε_i is that at the previous iteration; T denotes the reference deformation in the training data set, which can be acquired from previous operations or related database; t_i is the current calculated deformation; k represents the training rate. In our experiments the training process is terminated when the difference between t_i and T is less than 0.3 mm. Given the trained parameters, we are able to predict the soft tissue deformations in our simulation case studies.

Experiments and Comparison

We have developed a prototype craniofacial visualization and surgery simulation system incorporating the above techniques. In addition, we have also implemented some auxiliary functions such as 3D metric measurement. Here, we tested our system with three experimental case studies. The first is to test the accuracy of our points matching method and make a comparison with two other algorithms which are based on epipolar lines [21] and sequence matching [22]. Figure 7 shows the original X-ray pair. We placed 130 markers on the subject's right submaxilary. The matching results are shown in Fig. 8.

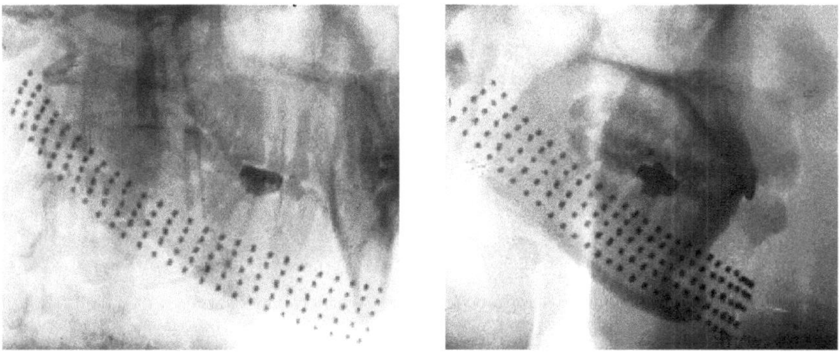

(a) (b)

Fig. 7. Original X-ray images pair: (**a**) 60° projection; (**b**) 90° projection.

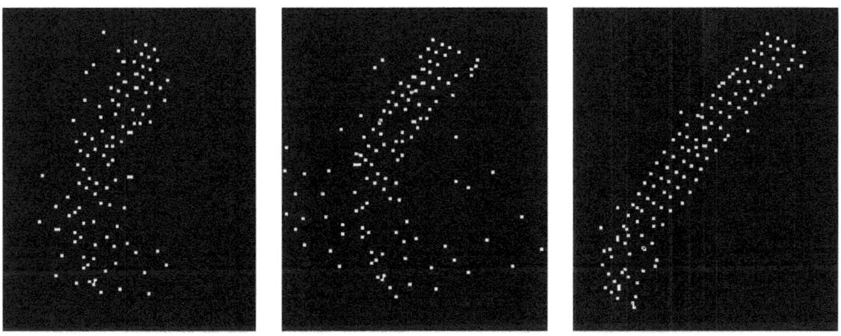

(a) (b) (c)

Fig. 8. Results of corresponding points matching based on different algorithms: (**a**) 3D points matched by epipolar lines searching; (**b**) 3D points matched by sequence matching algorithm; (**c**) 3D points matched by our method.

The 3D points in Fig. 8a, based on epipolar line searches, show clear matching errors. The reason is that the feature points are so dense that the error range is beyond the distance between two adjacent rows of feature points. In Fig. 8b, as the limited adaptability of sequence matching algorithm, a number of points are badly mismatched. The result would be good if all points satisfy the ordering constraints, which is too strict to be satisfied. The result in Fig. 8c, which is from our proposed matching method, demonstrates the best performance. In optimization searching by evolutionary programming, the generation scale number is $\frac{130(130-1)}{4} \approx 5000$. The time cost in the whole process is 2.7 s and the accuracy of matching is 93.8%.

The second experiment is to test the accuracy of the 3D reconstruction technique. We compare the results with the data measured by a helical CT. Table 1 lists the results of some vertical items measured by our prototype system including Euclidean distance and geodesic distances in comparison with the CT imaging data. N, Gn, Sto, Sn stand for some facial feature points defined by the surgeon.

Table 1. 3D measurement errors compared with CT (mm).

Feature points	Our system	CT	Difference	Permitted error tolerance
N-Gn	120.0	121.3	–1.3	±2.7
N-Sto	86.0	83.3	2.7	±2.9
Sto-Gn	35.42	38.4	–1.98	±2.2
Sn-Go	60.69	60.9	–0.21	±2.0
N-Sn	60.14	60.3	–0.16	±2.6
Sn-Sto	26.33	24.8	1.58	±2.6

Table 1 suggests that our reconstruction technique and the developed system offer good modeling accuracy. All measurements are within the required tolerances. For certain items, the developed technique is almost as accurate as CT scans, which is encouraging. However, our technique suffers from some limitations. Because it formulates the face model using lead markers, it is not suitable for regions where lead markers cannot be placed, for example, the areas covered by hair. The soft tissue thickness measured by M-ultrasound is another source of inaccuracy.

The third experiment is designed to investigate the effectiveness and precision of the soft tissue deformation technique and our parameter estimation algorithm. The patient underwent two osteotomy operations, Lefort I and mandible vertical osteotomy. In this experiment, the reference deformation data of Lefort I were taken, once the operation was complete and the measured data were used to train the Poisson's ratio (υ) for our double-layered finite element model. The trained parameter was then used to compute (simulate) the deformations incurred from the subsequent mandible vertical osteotomy. In our experiments, we fixed the value of the modulus of elasticity and only trained the Poisson's ratio, and we found one parameter sufficed. However, if both parameters need training, it is best to train them separately, i.e. to fix one while the other is being trained. Iterate the process until a predefined termination condition is met.

Using the trained Poisson's ratio υ=0.463, we ran our FEM simulation for the mandible vertical osteotomy. Figure 9 demonstrates the simulation results.

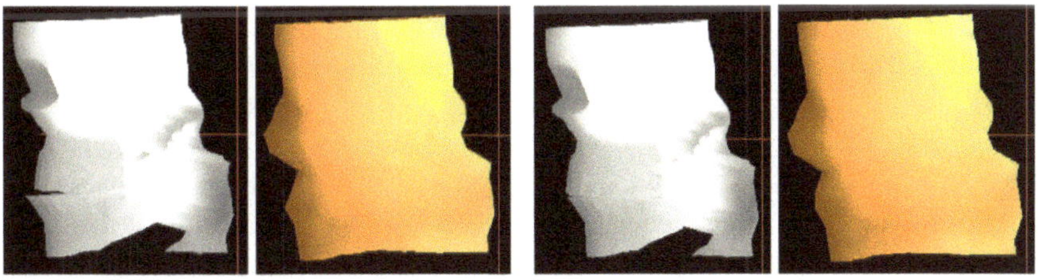

Fig. 9. Mandible vertical osteotomy simulation: (**a**) bone tissue before surgery; (**b**) soft tissue before surgery; (**c**) bone tissue after surgery; (**d**) soft tissue after surgery.

Table 2. Deformations in the mandible vertical osteotomy.

Regions	Fang teeth	Jawbone
Actual result	1.25:1	1.13:1
Simulation result	1.27:1	1.21:1

Table 2 shows the comparative results. According to the medical convention, the deformations are represented as a ratio of the bone tissue displacement and the soft tissue deformation. The comparison suggests that our simulation model coupled with the parameter estimation technique is capable of producing simulation results of good accuracy when both training and testing data are taken from a local region (e.g. the fang teeth region in Table 2). However, simulation errors may increase if the testing area is too far from the area where the training data are taken (e.g. the jawbone region).

Conclusions

We have presented an X-ray-based craniofacial visualization and surgery simulation system using only three X-ray photos. Because X-ray imaging records the tissue density rather than its 3D coordinates, a small number of X-ray images do not give sufficient information for the reconstruction of the cranium of a subject. Unlike the traditional CT-based approach, where the geometry is derived by evaluating the densities of the voxels, our first step is to formulate the face geometry using computer vision techniques. To reconstruct the skull, we measure the soft tissue thickness using M-ultrasound scanning data calibrated by front, lateral X-ray photos and subtract the tissue thickness from the face model.

In the step of corresponding points matching, we have developed a new algorithm by considering the similarity of the position sequences of the feature points. Our experimenting results show that our algorithm is able to overcome the difficulties of the existing techniques encountered with X-ray images.

The derived 3D geometry of the skull is valuable both for the simulation of soft tissue deformations and for the accurate measurement and evaluation of surgical results. Based on the obtained geometry of the face and the skull of a subject, we have implemented a craniofacial simulation technique using a double-layered finite element model for the computation of soft tissue deformations. In particular, we have proposed a simple yet effective algorithm for the estimation of the material parameters, including Young's modulus and Poisson's ratio. This has proven crucial to the accuracy of the simulation.

Compared with the CT-based craniofacial visualization approach, the primary merits of our technique are low radiation and low cost. It uses only ordinary medical equipment and therefore can be easily adopted in poorer and underprivileged countries and remote regions. We have conducted two surgical experiments, Lefort I and mandible vertical osteotomy. Practical measurements have shown that our craniofacial reconstruction technique produces small geometric errors, well within the required tolerance. The experiments also verified that our simulation technique, together with the parameter estimation algorithm, was of good simulation accuracy.

However, our development is at an early stage and has a number of limitations. The most obvious is that our current method for placing the lead markers is quite awkward and user-unfriendly. Nevertheless, our objective of this research is mainly to prove the research concept. The second is that it is not suitable for the areas that cannot be well represented by the lead markers, such as the places covered with hair. Therefore, it should not be considered to replace the CT system. The third disadvantage is that the simulation error may increase if the testing area is too far from the area where the training data are taken. But this is true for other existing simulation techniques as well.

The feedback from the doctors who participated in the experiments believed our method had unique advantages and was worthy of further development. In future work, we will devise a new easy-to-use marker placement method to cover the patient's face. Our idea is to design a special face mask to let the markers touch the face automatically, which can also guide the ultrasound scan with accurate placement of the probe. Using the developed prototype system, we also plan to investigate the relationship of the stiffness parameter values with human age, race, gender, ethnic origins and some other related factors.

References

1. Fu W (2002) Orthodontics. People Hygiene Press, Beijing
2. Zhang C (2003) Engineering and Clinic: Modern Medical Imaging. Science and Technology Press, Shanghai
3. Fuchs J, Kacherie M (2000) System performance of multi-slice computed tomography. IEEE Engineering in Medicine and Biology 15(3):375–377
4. Jarjan Z, Pozzi J, et al. (1995) Optimization of scanning and processing parameters for the 3D reconstruction in computerized tomography of the facial bones. Radiology Medicine 89(5):78–85

5. Adriana C, Joseph L, et al. (2003) Craniofacial applications of 3D laser surface scanning. The Journal of Craniofacial Surgery 14(4):449–455
6. Ki L, Hyeon H, Sean C, et al. (2007) Effect of cephalometer misalignment on calculations of facial asymmetry. Orthodontics and Dentofacial Orthopedics 132(1):15–27
7. Aoki Y, et al. (2001) Simulation of postoperative 3D facial morphology using a physics-based head model. The Visual Computer 17(2):121–131
8. Nakasima A, et al. (2005) Three-dimensional computer-generated head model reconstructed from cephalograms, facial photographs, and dental cast models. American Journal of Orthodontics and Dentofacial Orthopedics 127(3):282–292
9. Koch M, Gross H, Carls R, Buren F, et al. (1996) Simulation facial surgery using finite element models. In: Proceedings of SIGGRAPH 1996 Conference, ACM Press, pp. 125–132
10. Gladilin E, Zachow S, Deuflhard P, Hege H (2001) A biomechanical model for soft tissue simulation craniofacial surgery. In: International Workshop on Medical Imaging and Augmented Reality (MIAR 01), pp. 137
11. Duysak A, Zhang J, Ilankovan V (2003) Efficient modelling and simulation of soft tissue deformation using mass-spring systems. In: Proceedings of the 17th International Congress and Exhibition on Computer Assisted Radiology and Surgery (CARS 2003), London, pp. 123–127
12. Gibson (1997) 3D Chainmail: A fast algorithm for deforming volumetric objects. In: Proceedings of Interactive 3D Graphics Symposium 1997, pp. 149–154
13. Vukooyami (2001) A useful method in the cranium measurement by X-rays. Orthodontics Study 12:1–2
14. Coppini R (2002) 3D knowledge driven reconstruction of coronary trees. Medicine & Biology Engineering and Compute 29:535
15. Vladimir K (2003) Multi-camera scene reconstruction via graph cuts. PhD dissertation, Computer Science Department, Cornell University, Ithaca, New York, USA
16. Sylvain P, Francois X (2003) A surface reconstruction method using global graphic cut optimization. In: European Conference in Computer Vision, Rhodes, Greece, pp. 1–7
17. Pan J, Zhang Y (2005) Corresponding points matching based on position similarity. In: Proceedings of International Conference on Computer Graphics, Imaging and Vision (CGIV05), Beijing, pp. 33–37
18. Enciso R, Alexandroni S, Benyamein K, et al. (2004) Precision, repeatability and validation of indirect 3D anthropometric measurements with light-based imaging techniques. Biomedical Imaging: Macro to Nano. IEEE International Symposium 2004, pp. 1162–1167
19. Chai W, Sheng H, Sheng L, et al. (2001) Facial surgery simulation using finite element modelling. In: Proc. SPIE 2001, Vol. 4319, pp. 608–614
20. Jin H, Xiaohan S, Xin L, et al. (2006) Geometrically-based potential energy for simulating deformable objects. In: Pacific graphics 2006, Taipei, pp. 115–122
21. Gruen W (2001) Adaptive least squares correlation: A powerful image matching technique. In: Proceedings of International Conference of Remote Sensing 2001. Phoenix, pp. 65–71
22. Jia Y (2000) Computer Vision. Science Press, Beijing

OMOGENIA: A Semantically Driven Collaborative Environment

Aggelos Liapis[1]

Abstract. Ontology creation can be thought of as a social procedure. Indeed the concepts involved in general need to be elicited from communities of domain experts and end-users by teams of knowledge engineers. Many problems in ontology creation appear to resemble certain problems in software design, particularly with respect to the setup of collaborative systems. For instance, the resolution of conceptual conflicts between formalized ontologies is a major engineering problem as ontologies move into widespread use on the semantic web. Such conflict resolution often requires human collaboration and cannot be achieved by automated methods with the exception of simple cases. In this chapter we discuss research in the field of computer-supported cooperative work (CSCW) that focuses on classification and which throws light on ontology building. Furthermore, we present a semantically driven collaborative environment called OMOGENIA as a natural way to display and examine the structure of an evolving ontology in a collaborative setting.

Introduction

Computer-supported cooperative work (CSCW) is an area now re-emerging as an application of semantics, focusing on suitable forms of meaningful cooperation and collaboration between users or groups of users to perform a common task. It is concerned with design implementation and realization of computer support for cooperation to achieve the common goals [1]. In general, CSCW supports a range of applications such as shared editors, audio/video conferencing, computerized meeting rooms, group design tools, co-authoring systems, shared calendars, workflow systems, voting tools, whiteboard, and message-based conferencing [1]. In first-generation CSCW environments, knowledge sharing between its applications is unfortunately not taken to its full potential yet, implying a limited production of creativity and innovation. Ontologies are instrumental here, as they provide the necessary semantic resources of the world under discussion, to perform meaningful

[1] Vrije Universiteit Brussel – STARLab, Brussels, Belgium
liapis.aggelos@vub.ac.be

N. Magnenat-Thalmann et al. (eds.), *Recent Advances in the 3D Physiological Human*,
DOI 10.1007/978-1-84882-565-9_14, © Springer-Verlag London Limited 2009

annotation of application objects with shared concepts. Current collaborative environments do not provide users with content annotation mechanisms.

A new generation of semantically driven CSCW systems is required in which meaningful collaboration spans a wide range of synchronous, asynchronous, and semi-synchronous knowledge exchanges. Such systems need to be tailored in order to support a variety of collaborative users based on the context of their expertise [7]. Eventually, semantic interoperability may be achieved among heterogeneous and autonomous CSCW systems running locally or remotely at different platforms that may support both synchronous and asynchronous activity [7]. The following are some reasons for the usefulness of semantically driven CSCW systems:

- Support activities and share resources: Users need to communicate with each other in order to support their activities and share knowledge resources.
- User preferences: Current CSCW systems are heterogeneous and each offers a unique set of benefits. Users may be using their preferable system for long and they do not wish to give it up and adopt a new system. This challenge can be resolved with the combination of a user-friendly and fully customizable interface along with the implementation of a legacy system integration mechanism. A feature like this could make the transition to the new system trivial.
- Avoid the mistakes of the past: Current CSCW systems are typically designed from scratch repeating the principles of the past. This incurs significant delays, inconvenience, and involves a high risk of "reinventing the wheel" creating a clone of the old system.
- Improve efficiency and enhance functionality: CSCW systems are heterogeneous and each system offers limited functionality.

Related Work

The problem for integration of CSCW systems was first identified in the early 1990s and led to the following contributions.

Dewan [4] addresses some of the basic issues in interoperating heterogeneous collaborative systems, including coupling, semantics, and some implementation issues. This work mainly focuses on the integration of floor control mechanism with locking system. The results of this work show that it is possible to interoperate a synchronously coupled, fully replicated floor control system with a flexibly coupled, partially centralized lock system. Floor control is the simplest form of concurrency control, which allows only one user to input to the system at any given time. The user who wishes to operate the system has to request for floor control. The floor will be granted if it is free; otherwise request will be discarded or queued. Different techniques are in use such as turn-taking protocols. The problem with this technique is that it does not allow multiple users to perform actions in parallel even if their actions do not conflict. Lock-based concurrency control has addressed these problems. It allows users to obtain locks and work concurrently as long as they do not wish to work on the same objects. This work adopts an approach that assumes the source code and internal knowledge of the groupware applications to be interoperated are known [4].

Liapis [8] addresses the problem called intelligent collaboration transparency (ICT), in which the issues of interoperability between single-user heterogeneous applications are addressed. This work adopts a "black box" assumption, which assumes that the source code and internal knowledge of the groupware applications to be interoperated are unknown and no modification is allowed. Using this approach, the applications sharing infrastructure is interposed between the applications to be shared and their window environment at each site. Users can collaborate on the common task using their favorite single-user heterogeneous applications. The infrastructure captures and replays user input to the applications [8].

La Marca et al. [7] provide support for content as well as for coordination in collaborative work. It considers coordination and collaborative functionality as an aspect of the collaborative artifact rather than a collaborative application. Basically, this work considers coordination and collaboration as separated and independent of applications. This approach provides a mechanism to monitor the application access to a shared data repository and trigger user-supplied programs when interesting operations are performed. It enables heterogeneous single-user applications to be interoperated and converted into groupware without modification. This approach is limited and not referred to as application sharing systems [7].

None of the above approaches provides only partial integration solutions [6]. This chapter contributes to the state of the art by introducing a framework that enhances the potential for full integration [7]. To this end, we leverage the architecture of a CSCW system with an ontological model, a coordination model, and a user interface model. The description of these models is given below:

- The ontological model specifies all objects in the application, their relationships, and terminologies.
- The coordination model specifies how interactions occur within the system and describes workflow.
- The user interface model describes how the users see the system and how the system is presented at an interface level.

In a fully integrated system all three aspects would need to be integrated. From an architectural viewpoint we have the following levels of integration [9]:

- Integrated ontological level;
- Integrated coordination level;
- Integrated user interface level.

The concept of the security model and the transaction model is omitted in the above description. An integrative methodology is proposed based on this framework. This methodology involves decomposing the components of applications in order to fully understand the underlying concepts and analyzing them [9]. It supports different levels of integration including ontology, security, co-ordination, transaction, and user interface. Furthermore, it emphasizes the structural and terminological transformation as well as encoding and decoding in order to achieve different levels of integration.

The methodology consists of five steps:

- selection of applications (which are based on either same model or different model);
- analysis of applications (in terms of the ontological model, the security model, the coordination model, the transaction model, and the user interface model);
- finding common concepts (and resolve conflicts between the concepts);
- explanation of context and implicit concepts;
- mapping specification (at three levels: structural transformation, terminological transformation, and encoding and decoding).

What Is Ontology?

Ontologies are key instruments to represent semantics for information systems, and over the past decade they have received a wide variety of definitions and descriptions [5, 11, 13]. In essence, ontologies capture a relevant part of the semantics of a domain through concepts and their interrelations. Semantics can be understood in two ways [11]:

- Real-world semantics, which describe real-world things (e.g., a specific pallet with products), usually using some vocabulary.
- Formal semantics (e.g., constraints), which enable machine processing, such as reasoning and querying.

As communication enabler, it is clear that ontologies must reflect consensual agreements between communication partners. As such, it is essential to take collaboration into account when developing, maintaining, or reusing ontologies [5].

Bridging the Gap: The Role of Ontology in CSCW

The scenario approaches have been introduced in the HCI and CSCW communities to fill the gap that the "traditional approach" to design created by imposing a technological orientation, abstraction, and other "user-distant" features [9].

Scenarios can be developed through direct observation of users performing tasks in their work environment or through abstractions from theories of human activities [9]. By exploiting the scenario approaches of the HCI and CSCW communities, our aim was to balance technology orientation, which is well established in the ontology engineering community with user orientation, which is recommended by the HCI and CSCW communities. More specifically, we have tried to balance formality, which is a strong standard within the ontology engineering community, with informality, which is a HCI and CSCW requirement for not losing touch with the user [9].

Scenarios are a meaningful way of accounting for users' needs. They embody properties, qualities, or criteria that must be "put" in the system so that the system is accepted by its intended users [9]. Scenarios embody criteria that must be found when assessing the system. Scenarios are both requirements and assessment scenarios. This two-sided aspect of scenarios would need to be systemized.

A frequently occurring issue in collaborative ontology engineering is the exchange of basic information between stakeholders (e.g., between domain experts and knowledge engineers). An ontology should never be created without a basis to start from, and there are always materials (e.g., excel files, descriptive text documents, manuals, web pages) available that provide a lot of domain-specific information [10]. In current approaches and tools, these documents are not of second importance, and thus there is little support to handle them and their exchange. Often, the problem is resolved in an ad hoc manner, for instance by mailing them as attachments [10].

Another difficulty present in collaborative ontology engineering is meaning negotiation or the process of coming to a consensus [8]. While there is a lot of support for the divergent aspects of ontology engineering (e.g., brainstorming tools), the more difficult convergent phase is less supported by tools [8].

An environment to handle this part of the process should provide synchronous and asynchronous communication mechanisms through which people can communicate freely, as well as more structured approaches to guide them to consensus (e.g., agenda generator, conflict overviews) [9].

The Methodology: DOGMA

For our semantically driven CSCW system, we adopt and extend the DOGMA framework for ontology engineering. A DOGMA-inspired ontology is decomposed into a lexon base and a layer of ontological commitments [6]. A full formalization of DOGMA is given by De Leenheer et al. [2] and an overview is given by De Leenheer and Meersman [3].

A lexon base holds (multiple) intuitive conceptualization(s) of a particular domain. Each conceptualization is simplified to a "representation-less" set of context-specific binary fact types called lexons. A lexon represents a plausible binary fact type and is formally described as a 5-tuple $<V, term_1, role, co\text{-}role, term_2>$, where V is an abstract context identifier used to group logically related fact types in the conceptualization [3, 6].

Intuitively, a lexon may be read as within the context V, $term_1$ may have a relation with $term_2$ in which it plays a *role*, and conversely, in which $term_2$ plays a corresponding co-role [2, 12]. In this way, a lexon base can be described as a set of plausible elementary fact types.

Any application-dependent interpretation of a set of lexons is moved to a separate layer, called the commitment layer. This layer serves as a mediator between the plausible fact types (lexons) and their axiomatization in applications. Each commitment consists of a selection of appropriate lexons and a limitation on their use through the application-specific constraints [2, 12].

By separating the conceptualization (lexon base) from the axiomatization (commitment), this approach provides more common ground for reuse and agreements. For instance, a business rule stating that each person has exactly one address may hold for one partner's application, but may be too strong for another [6].

As with all things, information systems are in constant flux. Especially in a collaborative setting, where an ontology serves as a mediating instrument, we will find a high and continuous need for proper versioning and evolution management. DOGMA incorporates proper support for these issues through the use of change operators, change logs, and context dependencies. This includes detailed meta information such as the name of the contributor, time, date, and appropriate commentary on future changes or possible objections in a project [6].

The initial benefit is that DOGMA is an approach to ontology engineering, and as such it is not restricted to a specific ontology language such as OWL [6]. The second benefit is DOGMA's grounding in the linguistic representations of knowledge. This way, domain experts and knowledge engineers can use ordinary language constructs to communicate and capture knowledge [6, 2, 12]. The third DOGMA benefit is its strict separation between conceptualization (i.e., lexical representation of concepts and their relationships) and axiomatization (i.e., semantic constraints). This separation results in higher reuse, design scalability, and easier agreement procedures [3].

The Platform: OMOGENIA

OMOGENIA illustrated in Fig. 1 is based on the evolved version of two previous collaborative environments named "Virtual Designer" and "Hippocrates Collaborative Space" [9]. The particular version of the platform is supported by DOGMA to provide users with a semantically driven environment. The semantic aspect of the platform is being covered from the integration of a business process modeler, an ontology editor, and a series of text, image, and video annotation mechanisms, which are being described later in this chapter.

Fig. 1. The main interface of OMOGENIA.

The graphical user interface (GUI) of OMOGENIA suggests a highly customizable (see Fig. 1) linear process categorized in the following tabs.

The Tools of OMOGENIA

The following section describes in more detail the most significant tools and services of OMOGENIA focusing on their architecture and contribution on the particular context.

Brainstorming

The platform includes the following two brainstorming tools:

- Post-it notes
- Mind mapping

Both tools are using data mining and sorting algorithms to allow users access to relevant third-party online resources while brainstorming. In addition the post-it notes tool includes an integrated really simple syndication (RSS) reader dedicated to scan the web for the latest technologies, approaches, and events based on user input [7, 9, 10].

Collaborative Tools

Collaborative environments allow a virtual team or an organization to share their work, seeing what others are doing, commenting, and working together. We have integrated two synchronous and two asynchronous collaborative services [9]:

- Voice over Internet protocol (VoIP) via Skype communicator
- Remote access via virtual network computing (VNC)
- Semantic wiki

- Semantic repository (DOGMA)

Modeling Tools

The modeling category consists of an ontology builder named DOGMA workbench illustrated in Fig. 2.

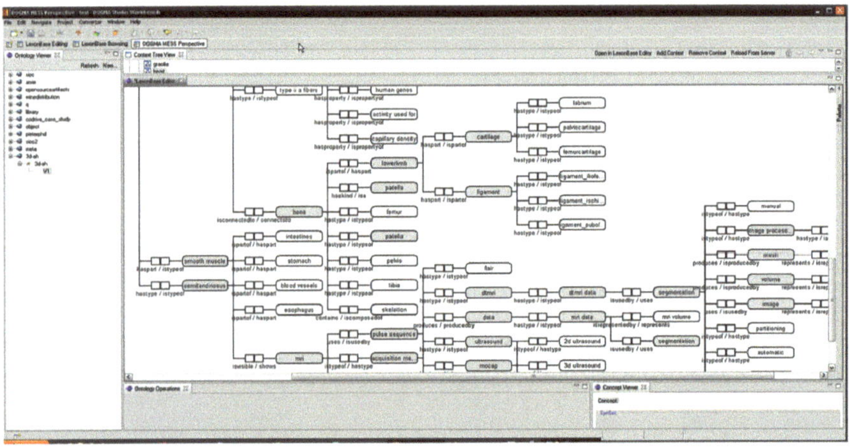

Fig. 2. DOGMA studio workbench – the modeling tool

The tool illustrated in Fig. 2 allows users to elicit and apply different ontological elements. Users can create, browse, and edit their ontologies as well as to keep a record of different versions and share selected elements with other partners [2, 3].

Recording Mechanisms

When a team collaborates using a groupware system its members must be coordinated in order to avoid possible conflicts [10]. Recording mechanisms allow participants to coordinate their work without communicating with others. This is an important characteristic as it prevents coordination breakdowns, which frequently occur during collaborative design [10]. The prototype consists of four different recording mechanisms to provide participants with a verification mechanism with regard to the evolution and security of the project. In addition participants can use the outputs as a reference for future work or to review the process [9]:

- Video and audio recording mechanism for synchronous collaborative sessions
- Automatic logging of users, activity when using the environment
- Automatic file versioning

- Collaborative history mechanism in the modeling tool

The outputs of these files are being saved in the file repository tool with appropriate meta information such as date, time, and participants.

Case Study

We illustrate with a realistic example from the Marie-Curie funded project named "Three Dimensional Anatomical Functional Models for the Human Musculoskeletal." The objective of this research and training network is to increase, by scientific exchange, the development of new technologies and knowledge around virtual representations of human body for interactive medical applications. The network has a specific goal: developing realistic functional three-dimensional models for the human musculoskeletal system, the methodology being demonstrated on the lower limb.

Description

The multidisciplinary nature of 3D-AH requires new collaboration approaches. The main challenge we are currently facing was that different groups while working in isolation were using different terminologies for the same concepts and methods. Communication becomes more difficult, unnecessary duplication of effort follows, and cross-fertilization between applications is reduced. The solution to this problem could be the collaborative creation of ontologies using the DOGMA methodology. Our main objective was to exploit the ontology in order to retrieve the associated information with the visualized structures. The information was retrieved using the Hippocrates collaborative space (HCS), the predecessor of OMOGENIA and the ontology wiki illustrated in Fig. 3.

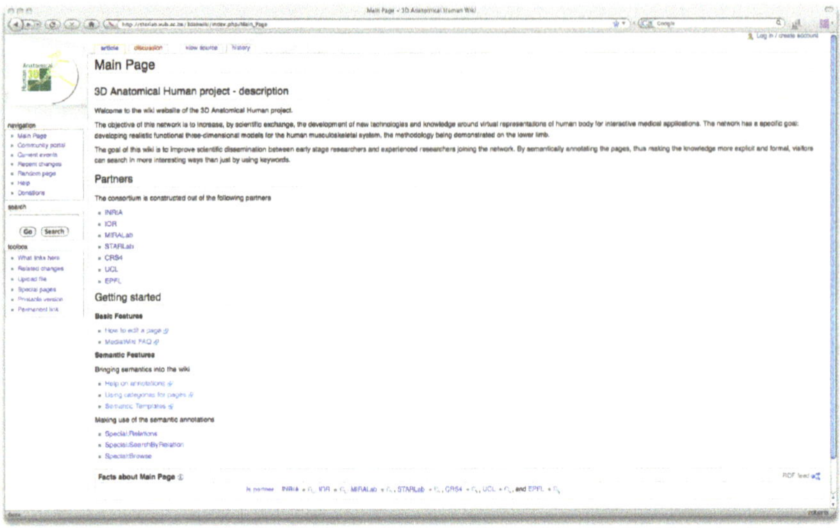

Fig. 3. The 3D-AH semantic media wiki

The semantic wiki allows some semantic links between concepts by using some rules. The wiki can be fed by other partners who can thus bring their knowledge into the application. In addition to the above collaborative tools, which allow specialists to share and model, different medical data sets are needed. For these reasons, we have developed an approach where experts in the domain of 3D modeling, image segmentation, MRI, and biomechanics are enabled to share data sets and interrelate them in an efficient and effective way. The process consists of the following steps:

- A set of domain concepts and relations mainly devoted (as in the traditional view on the ontology) to define properties of individuals, relations, and typical tasks involved in the application process.
- An image component including image semantic information (e.g., image sense and thematic descriptions for specific classes like muscle tissue) structured according to image segmentation methods and principles and modeled independently from the domain knowledge.
- A mapping between image and domain concepts

Once the ontology has been created using the DOGMA methodology and the collaborative platform, an interrogation mechanism was implemented and tested from the consortium in order to relate and compare existing ontologies in the field to the one created as part of the 3D-AH project.

To assist this process an API was defined to allow users to interrogate the ontology.

Current Results

The members of the 3D-AH consortium collaboratively created an initial version of the 3D-AH ontology during the collaboration days workshop organized by VUB-STARLab. After an extended brainstorming session using HCS brainstorming and data mining and capture services an automatic RDF export of the ontology was used in the semantic media wiki. For the elicitation of the lexons, participants used spreadsheets and imported them as ontology base in the DOGMA workbench. The use of spreadsheet for lexon elicitation from domain experts has been proven successful in other EC projects, including FF Poirot and CoDrive. This technique allows domain experts to express fact types in a controlled natural language.

Limitations

The results presented in the above sections illustrate an initial attempt of capturing the knowledge of the consortium. The intention is to combine the produced results with current state of the art and knowledge captured from domain experts within the context of 3D-AH.

Conclusions

The need for open CSCW systems has been discussed. To this end we have looked at generic models for CSCW focusing on their limitations and implemented our own model based on our previous work in the particular area.

Furthermore, we illustrated the features and the underpinning technology of a semantically driven collaborative platform named "OMOGENIA" focusing on the methodology that the particular tool follows (DOGMA).

We have also presented some initial results from the use of the particular environment in the context of 3D-AH for knowledge capturing and ontology engineering.

Acknowledgments The research performed in the context of this chapter is sponsored by the Marie Curie Research and Training Network's project 3D Anatomical Human (MRTN-CT-2006-035763). The author would like to thank Professor Robert Meersman for his guidance and support, the coordinator of the project Professor Nadia Magnenat Thalmann, the people involved in the workshop, and Christophe Debruyne and Margit Mikula for their assistance and contribution during the 3D-AH collaboration days workshop.

References

1. Benford S, Mariani J, Navarro L, Prinz W, Rodden T (1993) MOCCA: An Environment for CSCW Applications. In: ACM Organizational Computing Systems, California, Milpitas, pp. 172–177
2. De Leenheer P, De Moor A, Meersman R (2007) Context Dependency Management in Ontology Engineering: A Formal Approach. In: Journal on Data Semantics VIII, LNCS, vol. 4380, pp. 26–56, Springer
3. De Leenheer P, Meersman R (2007) Towards Community-Based Evolution of Knowledge-Intensive Systems. In: Proc. On The Move Federated Conferences: ODBASE (OTM 2007), LNCS, vol. 4803, pp. 989–1006, Springer
4. Dewan P. Sharma (1999) An Experiment in Interoperating Heterogeneous Collaborative Systems. In: Proc. of Sixth European Conference on Computer Supported Cooperative Work (ECSCW'99), Copenhagen, Denmark
5. Hepp M (2007) Possible Ontologies: How Reality Constrains the Development of Relevant Ontologies. IEEE Internet Computing 11(1):90–96
6. Hepp M, De Leenheer P, De Moor A, Sure Y (2008) Ontology Management for the Semantic Web, Semantic Web Services, and Business Applications, Springer-Verlag, New York
7. La Marca A, Edwards KW, Dourish P, Lamping J, Smith I, Thornton J (1999) Taking the Work out of Workflow: Mechanisms for Document-Centered Collaboration. In: Proc. of the 1999 European Conference on Computer Supported Cooperative Work (ECSCW '99)
8. Liapis A (2007) The Designer's Toolkit: A Collaborative Design Environment to Support Virtual Teams. In: Proceedings of IASK 2007 International Conference, Oporto, Portugal
9. Liapis A, Christiaens S, De Leenheer P (2008) Collaboration Across the Enterprise: An Approach for Enterprise Interoperability. In: International Conference on Enterprise Information (ICEIS 2008), Barcelona, Spain
10. Malins J, Liapis A (2007) The Design Educator's Toolkit: Virtual Environments in Art, Design and Education. Dublin Institute of Technology, Dublin, Ireland
11. Malins J, Watt S, Liapis A, McKillop C (2006) Tools and technology to support creativity in virtual teams In: S. MacGregor, T. Torres (Eds.), Virtual Teams and Creativity: Managing Virtual Teams Effectively for Higher Creativity, IGI Global, Harrisburg, PA
12. Meersman R (1999) The use of lexicons and other computer-linguistic tools in semantics, design and cooperation of database systems. In: Zhang, Y., Rusinkiewicz, M., Kambayashi, Y. (Eds.) Proceedings of the Conference on Cooperative Database Systems (CODAS 99), Springer-Verlag, pp. 1–14
13. Verheijen G, Van Bekkum J (1982) NIAM: An Information Analysis Method. In: Proc. of the IFIP TC-8 Conference on Comparative Review of Information System Methodologies (CRIS 82), North-Holland

Index

N. Magnenat-Thalmann et al. (eds.), *Recent Advances in the 3D Physiological Human*,
DOI 10.1007/978-1-84882-565-9_BM2,

FSC
www.fsc.org
MIX
Papier | Fördert
gute Waldnutzung
FSC® C083411

Kolibri 360 GmbH
Ferdinand-Jühlke-Straße 7
99095 Erfurt, Deutschland
produktsicherheit@kolibri360.de